Delmar's Test Preparation Series

Automobile Test

Manual Drive Train and Axles (Test A3)

2nd Edition

DELMAR
THOMSON LEARNING™

Australia Canada Mexico Singapore Spain United Kingdom United States

Delmar's Test Preparation Series
Automobile Test

Manual Drive Train and Axles (Test A3)
2nd Edition

Delmar Staff:

Business Unit Director:
Alar Elken
Executive Editor:
Sandy Clark
Acquisitions Editor:
Jack Erjavec
Team Assistant:
Bryan Viggiani
Developmental Editor:
Christopher Shortt

Executive Marketing Manager:
Maura Theriault
Executive Production Manager:
Mary Ellen Black
Production Manager:
Larry Main
Production Editor:
Betsy Hough
Production Editor:
Tom Stover

Channel Manager:
Mona Caron
Marketing Coordinator:
Brian McGrath
Cover Design:
Michael Egan
Cover Images Courtesy of:
DaimlerChrysler

Printed in Canada
3 4 5 6 7 8 9 XXX 05 04 03 02

For more information contact Delmar,
3 Columbia Circle, PO Box 15015,
Albany, NY 12212-5015.

Or find us on the World Wide Web at http://www.delmar.com or
http://www.autoed.com

ISBN: 0-7668-3426-3

Contents

Section 5 Sample Test for Practice

Section 6 Additional Test Questions for Practice

Section 7 Appendices

Preface

This book is just one of a comprehensive series designed to prepare technicians to take and pass every ASE test. Delmar's series covers all of the Automotive tests A1 through A8 as well as Advanced Engine Performance L1 and Parts Specialist P2. The series also covers the five Collision Repair tests and the eight Medium/Heavy Duty truck tests.

Before any book in this series was written, Delmar staff met with and surveyed technicians and shop owners who have taken ASE tests and have used other preparatory materials. We found that they wanted, first and foremost, *lots* of practice tests and questions. Each book in our series contains a sample test and additional practice questions. You will be hard-pressed to find a test prep book with more questions for you to practice with. We have worked hard to ensure that these questions match the ASE style in types of questions, quantities, and level of difficulty.

Technicians also told us that they wanted to understand the ASE test and to have practical information about what they should expect. We have provided that as well, including a history of ASE and a section devoted to helping the technician "Take and Pass Every ASE Test" with case studies, test-taking strategies, and test formats.

Finally, techs wanted refresher information and references. Each of our books includes an overview section that is referenced to the task list. The complete task lists for each test appear in each book for the user's reference. There is also a complete glossary of terms for each booklet.

So whether you're looking for a sample test and a few extra questions to practice with or a complete introduction to ASE testing, with support for preparing thoroughly, this book series is an excellent answer.

We hope you benefit from this book and that you pass every ASE test you take!

Your comments, both positive and negative, are certainly encouraged! Please contact us at:

Automotive Editor
Delmar Publishers
3 Columbia Circle
Box 15015
Albany, NY 12212-5015

Resource List

Erjavec, Jack. *Classroom Manual for Manual Transmissions and Transaxles.* Today's Technician. Albany, New York: Delmar Publishers, 1995.

Erjavec, Jack. *Shop Manual for Manual Transmission and Transaxles.* Today's Technician. Albany, New York: Delmar Publishers, 1995.

Newton Kenneth; Steeds, William; Garret, Thomas Kenneth; *The Motor Vehicle.* Boston: Butterworth International, 1989.

South David; Dwiggins, Boyce. *Delmar's Automotive Dictionary.* Albany, New York: Delmar Publishers, 1997.

The History of ASE

History

Originally known as The National Institute for Automotive Service Excellence (NIASE), today's ASE was founded in 1972 as a non-profit, independent entity dedicated to improving the quality of automotive service and repair through the voluntary testing and certification of automotive technicians. Until that time, consumers had no way of distinguishing between competent and incompetent automotive mechanics. In the mid-1960s and early 1970s, efforts were made by several automotive industry affiliated associations to respond to this need. Though the associations were non-profit, many regarded certification test fees merely as a means of raising additional operating capital. Also, some associations, having a vested interest, produced test scores heavily weighted in the favor of its members.

From these efforts a new independent, non-profit association, the National Institute for Automotive Service Excellence (NIASE), was established. In early NIASE tests, Mechanic A, Mechanic B type questions were used. Over the years the trend has not changed, but in mid-1984 the term was changed to Technician A, Technician B to better emphasize sophistication of the skills needed to perform successfully in the modern motor vehicle industry. In certain tests the term used is Estimator A/B, Painter A/B, or Parts Specialist A/B. At about that same time, the logo was changed from "The Gear" to "The Blue Seal," and the organization adopted the acronym ASE for Automotive Service Excellence.

ASE

ASE's mission is to improve the quality of vehicle repair and service in the United States through the testing and certification of automotive repair technicians. Prospective candidates register for and take one or more of ASE's many exams.

Upon passing at least one exam and providing proof of two years of related work experience, the technician becomes ASE certified. A technician who passes a series of exams earns ASE Master Technician status. An automobile technician, for example, must pass eight exams for this recognition.

The exams, conducted twice a year at over seven hundred locations around the country, are administered by American College Testing (ACT). They stress real-world diagnostic and repair problems. Though a good knowledge of theory is helpful to the technician in answering many of the questions, there are no questions specifically on theory. Certification is valid for five years. To retain certification, the technician must be retested to renew his or her certificate.

The automotive consumer benefits because ASE certification is a valuable yardstick by which to measure the knowledge and skills of individual technicians, as well as their commitment to their chosen profession. It is also a tribute to the repair facility employing ASE certified technicians. ASE certified technicians are permitted to wear blue and white ASE shoulder insignia, referred to as the "Blue Seal of Excellence," and carry credentials listing their areas of expertise. Often employers display their technicians' credentials in the customer waiting area. Customers look for facilities that display ASE's Blue Seal of Excellence logo on outdoor signs, in the customer waiting area, in the telephone book (Yellow Pages), and in newspaper advertisements.

To become ASE certified, contact:

National Institute for Automotive Service Excellence
13505 Dulles Technology Drive
Herndon, VA 20171-3421

ASE Testing

Participating in an Automotive Service Excellence (ASE) voluntary certification program gives you a chance to show your customers that you have the "know-how" needed to work on today's modern vehicles. The ASE certification tests allow you to compare your skills and knowledge to the automotive service industry's standards for each specialty area.

If you are the "average" automotive technician taking this test, you are in your mid-thirties and have not attended school for about fifteen years. That means you probably have not taken a test in many years. Some of you, on the other hand, have attended college or taken postsecondary education courses and may be more familiar with taking tests and with test-taking strategies. There is, however, a difference in the ASE test you are preparing to take and the educational tests you may be accustomed to.

Who Writes the Questions?

The questions on all ASE tests are written by service industry experts familiar with all aspects of the subject area. ASE questions are entirely job-related and designed to test the skills that you need to know on the job.

The questions originate in an ASE "item-writing" workshop where service representatives from domestic and import automobile manufacturers, parts and equipment manufacturers, and vocational educators meet in a workshop setting to share their ideas and translate them into test questions. Each test question written by these experts is reviewed by all of the members of the group.

All of the questions are pretested and quality-checked in a nonscoring section of tests by a national sample of certifying technicians. The questions that meet ASE's high standards of accuracy and quality are then included in the scoring sections of future tests. Those questions that do not pass ASE's stringent test are sent back to the workshop or are discarded. ASE's tests are monitored by an independent proctor and are administered and machine-scored by an independent provider, American College Testing (ACT).

Objective Tests

A test is called an objective test if the same standards and conditions apply to everyone taking the test and there is only one correct answer to each question. Objective tests primarily measure your ability to recall information. A well-designed objective test can also test your ability to understand, analyze, interpret, and apply your knowledge. Objective tests include true-false, multiple choice, fill in the blank, and matching questions. ASE's tests consist exclusively of four-part multiple-choice objective questions.

Before beginning to take an objective test, quickly look over the test to determine the number of questions, but do not try to read through all of the questions. In an ASE test, there are usually between forty and eighty questions, depending on the subject. Read through each question before marking your answer. Answer the questions in the order they appear on the test. Leave the questions blank that you are not sure of and move on to the next question. You can return to those unanswered questions after you have finished the others. They may be easier to answer at a later time after your mind has had additional time to consider them on a subconscious level. In addition, you might find information in other questions that will help you to answer some of them.

Do not be obsessed by the apparent pattern of responses. For example, do not be influenced by a pattern like **d, c, b, a, d, c, b, a** on an ASE test.

There is also a lot of folk wisdom about taking objective tests. For example, there are those who would advise you to avoid response options that use certain words such as *all, none, always, never, must,* and *only,* to name a few. This, they claim, is because nothing in life is exclusive. They would advise you to choose response options that use words that allow for some exception, such as *sometimes, frequently, rarely, often, usually, seldom,* and *normally.* They would also advise you to avoid the first and last option (A and D) because test writers, they feel, are more comfortable if they put the correct answer in the middle (B and C) of the choices. Another recommendation often offered is to select the option that is either shorter or longer than the other three choices because it is more likely to be correct. Some would advise you to never change an answer since your first intuition is usually correct.

Although there may be a grain of truth in this folk wisdom, ASE test writers try to avoid them and so should you. There are just as many **A** answers as there are **B** answers, just as many **D** answers as **C** answers. As a matter of fact, ASE tries to balance the answers at about 25 percent per choice **A, B, C,** and **D.** There is no intention to use "tricky" words, such as outlined above. Put no credence in the opposing words "sometimes" and "never," for example.

Multiple-choice tests are sometimes challenging because there are often several choices that may seem possible, and it may be difficult to decide on the correct choice. The best strategy, in this case, is to first determine the correct answer before looking at the options. If you see the answer you decided on, you should still examine the options to make sure that none seem more correct than yours. If you do not know or are not sure of the answer, read each option very carefully and try to eliminate those options that you know to be wrong. That way, you can often arrive at the correct choice through a process of elimination.

If you have gone through all of the test and you still do not know the answer to some of the questions, then guess. Yes, guess. You then have at least a 25 percent chance of being correct. If you leave the question blank, you have no chance. In ASE tests, there is no penalty for being wrong.

Preparing for the Exam

The main reason we have included so many sample and practice questions in this guide is, simply, to help you learn what you know and what you don't know. We recommend that you work your way through each question in this book. Before doing this, carefully look through Section 3; it contains a description and explanation of the questions you'll find in an ASE exam.

Once you know what the questions will look like, move to the sample test. After you have answered one of the sample questions (Section 5), read the explanation (Section 7) to the answer for that question. If you don't feel you understand the reasoning for the correct answer, go back and read the overview (Section 4) for the task that is related to that question. If you still don't feel you have a solid understanding of the material, identify a good source of information on the topic, such as a textbook, and do some more studying.

After you have completed the sample test, move to the additional questions (Section 6). This time answer the questions as if you were taking an actual test. Once you have answered all of the questions, grade your results using the answer key in Section 7. For every question that you gave a wrong answer to, study the explanations to the answers and/or the overview of the related task areas.

Here are some basic guidelines to follow while preparing for the exam:

- Focus your studies on those areas you are weak in.
- Be honest with yourself while determining if you understand something.
- Study often but in short periods of time.
- Remove yourself from all distractions while studying.
- Keep in mind the goal of studying is not just to pass the exam, the real goal is to learn!

During the Test

Mark your bubble sheet clearly and accurately. One of the biggest problems an adult faces in test-taking, it seems, is in placing an answer in the correct spot on a bubble sheet. Make certain that you mark your answer for, say, question 21, in the space on the bubble sheet designated for the answer for question 21. A correct response in the wrong bubble will probably be wrong. Remember, the answer sheet is machine scored and can only "read" what you have bubbled in. Also, do not bubble in two answers for the same question.

If you finish answering all of the questions on a test ahead of time, go back and review the answers of those questions that you were not sure of. You can often catch careless errors by using the remaining time to review your answers.

At practically every test, some technicians will invariably finish ahead of time and turn their papers in long before the final call. Do not let them distract or intimidate you. Either they knew too little and could not finish the test, or they were very self-confident and thought they knew it all. Perhaps they were trying to impress the proctor or other technicians about how much they know. Often you may hear them later talking about the information they knew all the while but forgot to respond on their answer sheet.

It is not wise to use less than the total amount of time that you are allotted for a test. If there are any doubts, take the time for review. Any product can usually be made better with some additional effort. A test is no exception. It is not necessary to turn in your test paper until you are told to do so.

Your Test Results!

You can gain a better perspective about tests if you know and understand how they are scored. ASE's tests are scored by American College Testing (ACT), a non-partial, non-biased organization having no vested interest in ASE or in the automotive industry. Each question carries the same weight as any other question. For example, if there are fifty questions, each is worth 2 percent of the total score. The passing grade is 70 percent. That means you must correctly answer thirty-five of the fifty questions to pass the test.

The test results can tell you:

- where your knowledge equals or exceeds that needed for competent performance, or
- where you might need more preparation.

The test results *cannot* tell you:

- how you compare with other technicians, or
- how many questions you answered correctly.

Your ASE test score report will show the number of correct answers you got in each of the content areas. These numbers provide information about your performance in each area of the test. However, because there may be a different number of questions in each area of the test, a high percentage of correct answers in an area with few questions may not offset a low percentage in an area with many questions.

It may be noted that one does not "fail" an ASE test. The technician who does not pass is simply told "More Preparation Needed." Though large differences in percentages may indicate problem areas, it is important to consider how many questions were asked in each area. Since each test evaluates all phases of the work involved in a service specialty, you should be prepared in each area. A low score in one area could keep you from passing an entire test.

There is no such thing as average. You cannot determine your overall test score by adding the percentages given for each task area and dividing by the number of areas. It doesn't work that way because there generally are not the same number of questions in each task area. A task area with twenty questions, for example, counts more toward your total score than a task area with ten questions.

Your test report should give you a good picture of your results and a better understanding of your task areas of strength and weakness.

If you fail to pass the test, you may take it again at any time it is scheduled to be administered. You are the only one who will receive your test score. Test scores will not be given over the telephone by ASE nor will they be released to anyone without your written permission.

3 Types of Questions on an ASE Exam

ASE certification tests are often thought of as being tricky. They may seem to be tricky if you do not completely understand what is being asked. The following examples will help you recognize certain types of ASE questions and avoid common errors.

Each test is made up of forty to eighty multiple-choice questions. Multiple-choice questions are an efficient way to test knowledge. To answer them correctly, you must think about each choice as a possibility, and then choose the one that best answers the question. To do this, read each word of the question carefully. Do not assume you know what the question is about until you have finished reading it.

About 10 percent of the questions on an actual ASE exam will use an illustration. These drawings contain the information needed to correctly answer the question. The illustration must be studied carefully before attempting to answer the question. Often, techs look at the possible answers then try to match up the answers with the drawing. Always do the opposite; match the drawing to the answers. When the illustration is showing an electrical schematic or another system in detail, look over the system and try to figure out how the system works before you look at the question and the possible answers.

Multiple-Choice Questions

One type of multiple-choice question has three wrong answers and one correct answer. The wrong answers, however, may be almost correct, so be careful not to jump at the first answer that seems to be correct. If all the answers seem to be correct, choose the answer that is the most correct. If you readily know the answer, this kind of question does not present a problem. If you are unsure of the answer, analyze the question and the answers. For example:

A rocker panel is a structural member of which vehicle construction type?

A. Front-wheel drive

B. Pickup truck

C. Unibody

D. Full-frame

Analysis:

This question asks for a specific answer. By carefully reading the question, you will find that it asks for a construction type that uses the rocker panel as a structural part of the vehicle.

Answer A is wrong. Front-wheel drive is not a vehicle construction type.

Answer B is wrong. A pickup truck is not a type of vehicle construction.

Answer C is correct. Unibody design creates structural integrity by welding parts together, such as the rocker panels, but does not require exterior cosmetic panels installed for full strength.

Answer D is wrong. Full-frame describes a body-over-frame construction type that relies on the frame assembly for structural integrity.

Therefore, the correct answer is C. If the question was read quickly and the words "construction type" were passed over, answer A may have been selected.

EXCEPT Questions

Another type of question used on ASE tests has answers that are all correct except one. The correct answer for this type of question is the answer that is wrong. The word "EXCEPT" will always be in capital letters. You must identify which of the choices is the wrong answer. If you read quickly through the question, you may overlook what the question is asking and answer the question with the first correct statement. This will make your answer wrong. An example of this type of question and the analysis is as follows:

All of the following are tools for the analysis of structural damage EXCEPT:

A. height gauge.

B. tape measure.

C. dial indicator.

D. tram gauge.

Analysis:

The question really requires you to identify the tool that is not used for analyzing structural damage. All tools given in the choices are used for analyzing structural damage except one. This question presents two basic problems for the test-taker who reads through the question too quickly. It may be possible to read over the word "EXCEPT" in the question or not think about which type of damage analysis would use answer C. In either case, the correct answer may not be selected. To correctly answer this question, you should know what tools are used for the analysis of structural damage. If you cannot immediately recognize the incorrect tool, you should be able to identify it by analyzing the other choices.

Answer A is wrong. A height gauge *may* be used to analyze structural damage.

Answer B is wrong. A tape measure may be used to analyze structural damage.

Answer C is correct. A dial indicator may be used as a damage analysis tool for moving parts, such as wheels, wheel hubs, and axle shafts, but would not be used to measure structural damage.

Answer D is wrong. A tram gauge *is* used to measure structural damage.

Technician A, Technician B Questions

The type of question that is most popularly associated with an ASE test is the "Technician A says. . . Technician B says. . . Who is right?" type. In this type of question, you must identify the correct statement or statements. To answer this type of question correctly, you must carefully read each technician's statement and judge it on its own merit to determine if the statement is true.

Typically, this type of question begins with a statement about some analysis or repair procedure. This is followed by two statements about the cause of the problem, proper inspection, identification, or repair choices. You are asked whether the first statement, the second statement, both statements, or neither statement is correct. Analyzing this type of question is a little easier than the other types because there are only two ideas to consider although there are still four choices for an answer.

Technician A, Technician B questions are really double true or false questions. The best way to analyze this kind of question is to consider each technician's statement separately. Ask yourself, is A true or false? Is B true or false? Then select your answer from the four choices. An important point to remember is that an ASE Technician A, Technician B question will never have Technician A and B directly disagreeing with each other. That is why you must evaluate each statement independently. An example of this type of question and the analysis of it follows.

Structural dimensions are being measured. Technician A says comparing measurements from one side to the other is enough to determine the damage. Technician

B says a tram gauge can be used when a tape measure cannot measure in a straight line from point to point. Who is right?

A. A only
B. B only
C. Both A and B
D. Neither A nor B

Analysis:

With some vehicles built asymmetrically, side-to-side measurements are not always equal. The manufacturer's specifications need to be verified with a dimension chart before reaching any conclusions about the structural damage.

Answer A is wrong. Technician A's statement is wrong. A tram gauge would provide a point-to-point measurement when a part, such as a strut tower or air cleaner, interrupts a direct line between the points.

Answer B is correct. Technician B is correct. A tram gauge can be used when a tape measure cannot be used to measure in a straight line from point to point.

Answer C is wrong. Since Technician A is not correct, C cannot be the correct answer.

Answer D is wrong. Since Technician B is correct, D cannot be the correct answer.

Most-Likely Questions

Most-likely questions are somewhat difficult because only one choice is correct while the other three choices are nearly correct. An example of a most-likely-cause question is as follows:

The most likely cause of reduced turbocharger boost pressure may be a:

A. westgate valve stuck closed.
B. westgate valve stuck open.
C. leaking westgate diaphragm.
D. disconnected westgate linkage.

Analysis:

Answer A is wrong. A westgate valve stuck closed increases turbocharger boost pressure.

Answer B is correct. A westgate valve stuck open decreases turbocharger boost pressure.

Answer C is wrong. A leaking westgate valve diaphragm increases turbocharger boost pressure.

Answer D is wrong. A disconnected westgate valve linkage will increase turbocharger boost pressure.

LEAST-Likely Questions

Notice that in most-likely questions there is no capitalization. This is not so with LEAST-likely type questions. For this type of question, look for the choice that would be the least likely cause of the described situation. Read the entire question carefully before choosing your answer. An example is as follows:

What is the LEAST likely cause of a bent pushrod?

A. Excessive engine speed
B. A sticking valve
C. Excessive valve guide clearance
D. A worn rocker arm stud

Analysis:
Answer A is wrong. Excessive engine speed may cause a bent pushrod.
Answer B is wrong. A sticking valve may cause a bent pushrod.
Answer C is correct. Excessive valve clearance will not generally cause a bent pushrod.
Answer D is wrong. A worn rocker arm stud may cause a bent pushrod.

Summary

There are no four-part multiple-choice ASE questions having "none of the above" or "all of the above" choices. ASE does not use other types of questions, such as fill-in-the-blank, completion, true-false, word-matching, or essay. ASE does not require you to draw diagrams or sketches. If a formula or chart is required to answer a question, it is provided for you. There are no ASE questions that require you to use a pocket calculator.

Testing Time Length

An ASE test session is four hours and fifteen minutes. You may attempt from one to a maximum of four tests in one session. It is recommended, however, that no more than a total of 225 questions be attempted at any test session. This will allow for just over one minute for each question.

Visitors are not permitted at any time. If you wish to leave the test room, for any reason, you must first ask permission. If you finish your test early and wish to leave, you are permitted to do so only during specified dismissal periods.

You should monitor your progress and set an arbitrary limit to how much time you will need for each question. This should be based on the number of questions you are attempting. It is suggested that you wear a watch because some facilities may not have a clock visible to all areas of the room.

4 An Overview of the System

Manual Drive Train and Axles (Test A3)

The following section includes the task areas and task lists for this test and a written overview of the topics covered in the test.

The task list describes the actual work you should be able to do as a technician that you will be tested on by the ASE. This is your key to the test and you should review this section carefully. We have based our sample test and additional questions upon these tasks, and the overview section will also support your understanding of the task list. ASE advises that the questions on the test may not equal the number of tasks listed; the task lists tell you what ASE expects you to know how to do and be ready to be tested upon.

At the end of each question in the Sample Test and Additional Test Questions sections, a letter and number will be used as a reference back to this section for additional study. Note the following example: **B.4.**

Task List

B. Transmission Diagnosis and Repair (6 Questions)

Task B.4 — Remove and replace transmission; inspect transmission mounts.

Example:

1. During manual transmission removal and replacement:
   A. the driveshaft may be installed in any position on the differential pinion gear flange.
   B. the transmission weight may be supported by the input shaft in the clutch disc hub.
   C. the engine support fixture should be installed after the transmission-to-engine bolts are loosened.
   D. the clutch disc must be aligned with an aligning tool before transmission installation. (B.4)

Analysis:

Question #1

Answer A is wrong. The driveshaft must be installed in the same position that it was removed from.

Answer B is wrong. The transmission cannot be supported by the input shaft, it could fall and damage the transmission or injure the technician.

Answer C is wrong. The engine support fixture should be installed when the transmission bolts are tight.

Answer D is correct.

Task List and Overview

A. Clutch Diagnosis and Repair (6 Questions)

Task A.1 Diagnose clutch noise, binding, slippage, pulsation, chatter, pedal feel/effort, and release problems; determine needed repairs.

Clutch chatter is felt normally when the pressure plate has made initial contact with the flywheel. Things that can cause a clutch to chatter, such as weak torsional springs, will not absorb the shock of the clutch when contact is made. Other causes of chatter could be a bent clutch disc or a burned or glazed lining on the disc. If the flywheel is glazed, has excessive runout, or is scored, clutch chatter may occur.

If a clutch was disengaged, the input shaft on the transmission would not be spinning, eliminating the bearing noise coming from the input shaft bearing. A pilot bearing would be a probable cause of noise in this case. With the input shaft not spinning, the pilot bearing is spinning on the end of the input shaft. This could cause a bearing noise due to a bad or worn pilot bearing.

Task A.2 Inspect, adjust, and replace clutch pedal linkage, cables, and automatic adjuster mechanisms, brackets, bushings, pivots, and springs.

When there is no clutch pedal free play, the clutch is not fully engaged. The release bearing is touching the fingers of the pressure plate. This will relieve some of the clamping on the clutch disc. This may cause the clutch to slip; it will not cause hard shifting, improper clutch release, or transaxle gear damage. These would all be signs of too much clutch pedal free play.

Many late-model vehicles have self-adjusting cables. The cable is adjusted when the clutch pedal is released as the clutch disc wears from normal use. These systems use a constant-running release bearing. It is always in contact with the pressure plate. The clutch pedal will not have any clutch pedal free play.

Task A.3 Inspect, adjust, replace, and bleed hydraulic clutch slave and master cylinders, lines, and hoses.

The hydraulic system for a hydraulic clutch is totally separate from the brake system. The clutch master cylinder is mounted to the fire wall in the engine compartment. A line runs from the clutch master cylinder to the slave cylinder on the bell housing. A rod connected to the clutch pedal goes through the fire wall into the clutch master cylinder. When the rod is pushed into the clutch master cylinder, it forces fluid through the line, which actuates the slave cylinder to release the clutch.

Air in the system will prevent the clutch from disengaging properly when the clutch pedal is fully depressed. Air is compressible and will compress in the hydraulic system before the spring pressure in the pressure plate releases the clutch disc from the flywheel. The conditions of less free play, worn facings, or a scored pressure plate will cause clutch problems, but will not affect the release of the clutch.

Task A.4 Inspect, adjust, and replace release (throwout) bearing, lever, and pivot.

On a clutch with an adjustable linkage, the release bearing should not be in contact with the pressure plate fingers. If the release bearing is not touching the fingers, it will not make any noise even if the bearing is bad. Clutch pedal free play is the distance between the release bearing and the pressure plate fingers. It is the gap or movement in the clutch pedal before the release bearing contacts the pressure plate and releases the clutch.

Hydraulic-controlled clutch systems use a release bearing that is always in contact with the fingers on the pressure plate. There is no manual adjustment on a hydraulic clutch system; it adjusts automatically as the clutch disc wears. When a clutch is disengaged, the release bearing moves towards the pressure plate. The release bearing continues to move towards the pressure plate fingers and compresses the springs in the pressure plate to release the clutch.

Task A.5 Inspect and replace clutch disc and pressure plate assembly.

Excessive crankshaft end play causes the pressure plate to move away from the clutch release bearing, which may result in improper clutch release. Loose engine main bearings may cause an oil leak at the rear main bearing, which contaminates the clutch facings with oil, resulting in clutch slippage. An improper pressure plate-to-flywheel position causes engine vibrations.

If the clutch facing is worn too thin, the clamping force of the pressure plate will not be as much as it was when the clutch disc was at full thickness. There will not be enough spring pressure left in the pressure plate to maintain a hard clamping force on the clutch disc because the springs in the pressure plate are fully extended and not applying enough pressure. The minimum, thickness of clutch disc lining is 0.012 inch (0.3 mm). Slippage will occur if the lining is any thinner.

Proper installation of the clutch disc is critical, the damper spring offset must face the transmission. The clutch disc is normally marked to indicate which side should face the flywheel. If the damper spring offset is toward the engine the springs may contact the flywheel or the flywheel bolts, damaging these components.

Task A.6 Inspect and replace pilot bearing.

When the clutch is engaged, the transmission input shaft rotates at the same speed as the engine flywheel and pilot bearing at all times. When the clutch is disengaged, the flywheel and pilot bearing rotate on the end of the transmission input shaft and turn faster than the shaft.

If a bushing-type pilot bearing is lubricated with bearing grease, friction actually will increase between the bushing and the transmission input shaft. Lubricate a bushing-type pilot bearing with SAE 30 motor oil. Lubricate a roller-type pilot bearing with wheel bearing grease.

Task A.7 Inspect and measure flywheel and ring gear; repair or replace as necessary.

If too much material is removed from the flywheel, the torsion springs on the clutch plate are moved closer to the mounting bolts on the flywheel, and these springs may contact the heads of the flywheel bolts. Removing excessive material from the flywheel moves the pressure plate forward, away from the release bearing. This action increases free play so the slave cylinder rod may not move far enough to release the clutch.

By not resurfacing the flywheel, damage or premature wear can be caused to the rest of the new assembly that was installed. A flywheel should be resurfaced every time a clutch assembly is replaced. Resurfacing the flywheel will ensure that it has the flatness and a microfinish it needs to ensure that the new clutch disc breaks in properly. If it is not resurfaced, the clutch disc will probably glaze and chatter. The flywheel should be cleaned with hot water and soap to remove all residue on the surface after it has been resurfaced.

Inspect the shaft tip that rides in the pilot bearing for smoothness. Check the splines of the shaft for any wear that could prevent the clutch disc from sliding evenly and smoothly. If the splines have excessive wear or damage, this will cause the clutch to engage roughly. A clutch disc that is bent or that has weak torsional springs will also cause the clutch to engage roughly.

Task A.8 Inspect engine block, clutch (bell) housing, and transmission case mating surfaces; determine needed repairs.

If the bell housing is not aligned properly with the engine block because of something being pinched between them or because a burr or imperfection on one of the mating surfaces, the clutch will not make even contact with the flywheel. This will cause the clutch to chatter and grab because of the uneven contact when the clutch is being engaged. Reduced clutch pedal free play, growling noises, or vibrations at high speeds are not symptoms of a misalignment condition.

If a clutch disc or an input shaft is bent due to careless removal or installation of the transmission, this will cause misalignment problems. If the transmission is misaligned when it is installed, this can cause the pilot bushing or bearing to wear out prematurely.

Task A.9 Measure flywheel-to-block runout and crankshaft end play; determine needed repairs.

To measure crankshaft end play, mount the magnetic dial indicator on the back of the engine block. Position the dial indicator to the flywheel. Push the flywheel toward the front of the engine until it stops. Adjust the dial indicator to zero and then pull the flywheel toward the back of the engine. The reading on the dial indicator will be the crankshaft end play.

The flywheel can be checked with the dial indicator setup after checking the crankshaft end play. Observe the dial indicator as the flywheel is rotated. The measured movement on the dial indicator is the runout and should be compared to the vehicle specifications.

Excessive main bearing wear will cause low oil pressure or a rear main oil leak, possibly causing the clutch disc to become contaminated with engine oil. A thrust bearing is placed between the crankshaft main bearing cap and the side of a crankshaft journal. These thrust bearings are put into place to control the forward and rearward movement of the crankshaft during acceleration and deceleration. The proper thickness thrust bearing is selected when the engine is assembled to set the crankshaft end play.

Task A.10 Measure clutch (bell) housing bore-to-crankshaft runout and face squareness; determine needed repairs.

To correct bell housing face runout, shims are sometimes installed between the engine block and the bell housing mating surfaces. Use a dial indicator to measure bell housing bore alignment. Bell housing bore misalignment may be corrected by turning eccentric dowels in the engine block at the bell housing mounting surface. Replacing the engine mounts, installing shims, or replacing the clutch will not compensate or correct this problem.

Task A.11 Inspect, replace, and align powertrain mounts.

Engine and transmission mounts should be inspected for broken, sagged, oil-soaked, or deteriorated conditions. Any of these mount conditions may cause a grabbing, binding clutch. On a rear-wheel drive car, damaged engine or transmission mounts may cause improper driveshaft angles, which results in a vibration that changes in intensity when the vehicle accelerates and decelerates.

On a vehicle with a manual clutch, part of the clutch linkage is connected from the frame of the vehicle to the engine block. If the vehicle has a broken engine mount, the engine can lift off the frame on acceleration. This can cause the clutch linkage to move and apply release pressure to the release bearing fork. This may cause the release bearing to apply release pressure on the pressure plate, causing the clutch disc to slip between the flywheel and the pressure plate.

B. Transmission Diagnosis and Repair (6 Questions)

Task B.1 Diagnose transmission noise, hard shifting, jumping out of gear, and fluid leakage problems; determine needed repairs.

A misaligned or loose transmission case or clutch housing may cause the transmission to slip out of gear. Check the mounting of the transmission to the engine block for looseness or dirt between the two cases. Broken or loose engine mounts will also cause misalignment. This problem may also be caused by insufficient spring tension of the shift rail detent spring or bent or worn shift forks, levers, or shafts. Other internal problems may cause this condition, such as a worn input shaft pilot bearing, bent output shaft, or a worn or broken synchronizer.

If the transmission shifts hard or the gears clash while shifting, a common cause of the problem is the clutch. Check the clutch pedal free-travel adjustment. Make sure the clutch releases completely. Also check for worn clutch parts and a binding input shaft pilot bearing. Shift linkage problems can also cause this problem. If the shift lever is worn, binding, or out of adjustment, proper engagement of gears is impossible. An unlubricated linkage will also cause shifting difficulties. Worn or damaged shift levers, rails, or forks can also cause this problem.

Task B.2 Inspect, adjust, and replace transmission external shifter assembly, shift linkages, brackets, bushings, grommets, pivots, and levers.

Most external shift linkages and cables require adjusting, and a similar adjustment procedure is used on some vehicles. Raise the vehicle and place the shifter in the neutral position to begin the shift linkage adjustment. With a lever-type shift linkage, install a rod in the adjustment hole in the shifter assembly. Adjust the shift linkages by loosening the rod retaining locknuts and moving the levers until the rod fully enters the alignment holes. Tighten the locknuts and check the shift operation in all gears.

Transmissions with internal linkage do not have provisions for adjustment. However, external linkages, both floor and column mount, can be adjusted. Linkages are adjusted at the factory, but worn parts may make adjustment necessary. Also, after a transmission has been disassembled, the shift lever may need adjustment.

Task B.3 Inspect and replace transmission gaskets, sealants, seals, and fasteners; inspect sealing surfaces.

Excessive output shaft or excessive input shaft end play results in lateral shaft movement that will not adversely affect the extension housing seal. A worn output shaft bearing will not cause premature extension housing seal failure. If the driveshaft yoke has a score or imperfection on the shaft, it could damage the seal and cause the transmission to leak fluid at the rear of the transmission. Replacing the seal will not correct this condition until the yoke is replaced.

A cork gasket should be installed as it is when it comes out of the box. It was made to be installed dry and does not require any type of added sealant to help the gasket seal any better. A spray adhesive sometimes may be used to hold the gasket in place to help installation. A rubber gasket should not use any additional gasket sealant when installed. It will become too slippery and may not position correctly when being installed.

A transmission mount absorbs a lot of torque and vibration in the rubber mount when shifting and accelerating. If the rubber mount becomes saturated with oil, the oil will deteriorate the rubber and weaken it. This will eventually cause the mount to fail. Oil will not cause the mount to crack, but the oil will make the rubber in the mount feel softer and spongy.

Task B.4 Remove and replace transmission; inspect transmission mounts.

The driveshaft should be installed in the original position on the rear axle pinion flange. If it is not installed in the original position, you may experience a driveline vibration. If the transmission is supported from the input shaft, the weight of the transmission itself could cause the input shaft and the clutch disc to be bent or damaged. The engine support fixture must be installed before the transmission-to-engine bolts are even loosened. When the transmission is being removed, the clutch disc may move, causing it to be misaligned. A clutch disc alignment tool must be used before the transmission is installed to align the clutch disc with the flywheel.

Task B.5 Disassemble and clean transmission components; reassemble transmission.

A synchronizer hub that does not slide smoothly over the blocker ring causes the hub to jam, resulting in hard shifting. The stronger hub and sleeve should be marked before disassembly so it can be installed the same way that it came apart.

Clean the transmission case with a steam cleaner, degreaser, or cleaning solvent. As you begin to disassemble the unit, pay close attention to the condition of its parts. Using a dial indicator, measure and record the end play of the input and main shafts. This information will be needed during assembly of the unit to select the appropriate selective shims and washers. The transmission parts should be cleaned with solvent before assembly.

Task B.6 Inspect and repair or replace transmission shift cover and internal shift forks, bushings, levers, shafts, sleeves, detent mechanisms, interlocks, and springs.

Shift rails should be inspected to be sure that they are not bent. A bent shift rail will not cause the transmission to jump out of gear or result in a gear clash or a gear noise. Hard shifting may be a cause of a bent shift rail. The shift rail is linked to the shifter handle and if the rail is bent, it may interfere with other parts in the transmission, causing the hard shift feeling when the shift lever is moved from certain gears.

A linkage that is bent or a linkage that is out of adjustment are both common causes of a transmission that does not shift properly. Failure to go into gear is more commonly caused by a broken shift fork than by a damaged gear.

The shift linkages internal to the transmission are located at the top or side of the housing. Mounted inside the transmission is the control end of the shifter and the shift controls. The shift controls consist of the shift rail and the shift fork. As the shift fork moves toward the preferred gear, it moves the synchronizer sleeve to lock the speed gear to the shaft.

Task B.7 Inspect and replace input (clutch) shaft, bearings, and retainers.

A worn pilot bearing contact area on the input shaft or a worn pilot bearing could result in noise with the clutch pedal depressed. The main shaft is not turning with the engine idling, the clutch released, and the transmission in neutral. Since the input shaft is not turning with the clutch released, a rough input shaft roller bearing or needle bearings would not result in a growling noise under this condition.

An input shaft only uses one type of bearing: a ball bearing located toward the front half of the shaft (normally a pressed fit to the shaft). The bearing is lubricated by the fluid in the transmission. A needle bearing would not support the load that an input shaft is subjected to.

Task B.8 Inspect and replace main shaft, gears, thrust washers, bearings, and retainers/snap rings.

A worn first-speed gear blocking ring or synchronizer sleeve may cause hard shifting but would not result in noise while driving in first gear. A worn, rough mainshaft bearing would cause a growling noise in all gears. Chipped and worn first-speed gear teeth would cause a growling noise in first gear.

The main shaft is not drilled with oil journals. Inspect the bearing surfaces of the main shaft; it should be smooth and show no signs of overheating. Also inspect the gear journal areas on the shaft for roughness, scoring, and other defects. Check the shaft splines for wear, burrs, and other conditions that would interfere with the slip yoke's ability to slide smoothly on the splines.

Task B.9 Inspect and replace synchronizer hub, sleeve, keys (inserts), springs, and blocking (synchronizing) rings; measure blocking ring clearance.

The blocking ring dog teeth tips should be pointed with smooth surfaces. Clearance between the blocking ring and the matching gear dog teeth is important for proper shifting. The synchronizer sleeve must slide freely on the synchronizer hub. The threads on the blocking ring in the cone area must be sharp to get a good bite on the gear to stop it from spinning and to make a synchronized non-clashing shift.

If the clearance between the blocking ring and the fourth-speed gear dog teeth is less than specified, the blocking ring is worn, which results in hard shifting. This problem would not result in noise while driving in fourth gear.

Task B.10 Inspect and replace counter (cluster) gear, shaft, bearings, thrust washers, and retainers/snap rings.

Since the countergear is turning with the clutch pedal released in neutral, and in gear, damaged countergear bearings may cause a growling noise with the engine idling with the transmission in neutral and the clutch pedal released. It will also cause a growling noise while driving in any gear.

All countergears should show wear patterns in the center of their teeth. These wear patterns should appear as a polished finish, with little wear on the gear face. Check the gears' teeth carefully for chips, pitting, cracks, or breakage. Also, inspect the bearing surfaces to make sure they are smooth. Any damage to the assembly requires replacement.

Task B.11 Inspect and replace reverse idler gear, shaft, bearings, thrust washers, and retainers/snap rings.

Since the reverse idler gear is only in mesh with reverse gear, this gear rotates in reverse gear only.

Inspect the reverse idler gear for pitted, cracked, nicked, or broken teeth. Check its center bore for a smooth surface. Carefully inspect the needle bearings on which the idler gear rides for wear, burrs, and other defects. Also, inspect the reverse idler gear's shaft surface for scoring, wear, and other imperfections. Replace any part that is damaged or excessively worn.

A worn extension housing bushing may cause premature extension housing seal wear and fluid leaks. Excessive mainshaft end play has no effect on speedometer operation.

Task B.12 Measure and adjust shaft, gear, and synchronizer end play.

In fourth gear, the 1-2 synchronizer is moved ahead so the synchronizer hub is meshed with the dog teeth on the fourth-speed gear on the input shaft. Excessive input shaft end play would cause the transmission to jump out of fourth gear.

While disassembling a transmission, the technician should be taking end play readings. These readings will be recorded, and when the transmission is reassembled, select-fit thrust washers and shims will be used to set all parts to specifications. A mainshaft should not be replaced unless, during inspection, excessive wear or damage is found. Bearings and snap rings can be reused if nothing is found during cleaning and inspection.

Task B.13 Measure and adjust bearing preload.

Bearing preload is adjustable. It is normally adjusted by placing force on the bearing using a select-fit shim behind the bearing or by an adjusting nut that applies pressure to the bearing. Bearing preload is used to take play out of the bearing so the bearing takes the load correctly. Packing a bearing is not part of setting bearing preload.

A worn speedometer drive and driven gears may cause erratic speedometer operation, but this problem would not result in premature extension housing bushing failure. A plugged transmission may cause excessive transmission pressure and fluid leaks, but this problem would not affect extension housing bushing wear. Excessive mainshaft end play causes lateral shaft movement, but this does not cause excessive extension housing bushing wear. Metal burrs between the extension housing and the transmission case cause misalignment of the extension housing, which forces the extension housing bushing against the driveshaft yoke in one location, resulting in bushing wear.

Task B.14 Inspect, repair, and replace extension housing and transmission case mating surfaces, bores, bushings, and vents.

Check the extension housing for cracks and repair or replace it as needed. Check the mating surfaces of the housing for burrs or gouges and file the surface flat. Inspect the speedometer cable or sensor for any leakage from the seal; replace the seal if fluid is leaking. Install a new gasket to the extension housing during installation. Check all threaded holes and repair any damaged bores with a thread repair kit. Check the bushing in the rear of the extension housing for excessive wear or damage. Always replace the rear extension housing seal.

Task B.15 Inspect and replace speedometer drive gear, driven gear, and retainers.

The speedometer driving gear is the gear located on the output shaft of the transmission and can be accessed by removing the rear extension housing. The speedometer driven gear is located on the vehicle speed sensor and is located in the rear extension housing. There are no speedometer gears located on the main shaft or the axle shaft.

Task B.16 Inspect, test, and replace transmission sensors and switches.

The backup lamp switch is normally located in the transmission on a manual transmission-equipped vehicle. The switch is normally open and has power going to it when the vehicle ignition is on. The backup lamp lights when the vehicle is shifted into reverse. The switch will close when the vehicle is shifted in reverse and provide a path to ground for the backup lamps to operate.

Task B.17 Inspect lubrication devices; check fluid level, and refill with proper fluid.

Transmission fluid level must be maintained at the level of the check plug in the transmission case or at a level marked on a transmission dipstick. Many late-model manual transmissions and transaxles use automatic transmission fluid (ATF) as a lubricant for reduced friction and improved vehicle fuel economy. Some manual transmissions use hypoid gear oil as a lubricant, and a few use motor oil.

The hypoid ring-and-pinion gearsets in rear-wheel driveaxles require hypoid gear oil, usually GL4 or GL5. Limited slip differentials require additional fluid additives. The viscosity of hypoid gear oil is higher than that of motor oil or ATF. It may be single-viscosity, such as SAE 90, or multiple-viscosity, such as 85W-90. Many final-drive gearsets in front-wheel drive (FWD) transaxles are not hypoid gears and use ATF or motor oil as a lubricant. Some FWD final drives are hypoid gearsets, however, and require GL4 or GL5 gear oil. Always follow the carmaker's specifications for fluid type, viscosity, and replacement intervals.

C. Transaxle Diagnosis and Repair (8 Questions)

Task C.1 Diagnose transaxle noise, hard shifting, jumping out of gear, and fluid leakage problems; determine needed repairs.

A worn fourth-speed synchronizer will only affect shifting in fourth gear. Excessive mainshaft end play may result in gear clash in all gears. A worn 3-4 shift fork may cause shifting problems in third and fourth gear, but this problem will not result in gear clash in all gears. A clutch disc sticking on the input shaft will cause the clutch not to release properly, resulting in gear clash in all gears.

A worn blocker ring, damaged speed gear, or a worn bushing could cause gear clash in a particular gear. Gear clash in all gears could be caused by stretched shifter cables.

Task C.2 Inspect, adjust, and replace transaxle external shift assemblies, linkages, brackets, bushings, grommets, cables, pivots, and levers.

A misadjusted shift linkage may cause many problems. If the linkage is misadjusted, the transmission may not be able to be shifted all the way into gear. This will cause further damage to the transmission. An improper shift linkage adjustment also may cause hard shifting or sticking in gear.

Most transmissions and transaxles are adjusted with the unit in neutral. A ¼-inch (6.35 mm) bar or drill bit is installed in the lever to hold the transaxle in neutral while the cables or linkage are adjusted. Transmissions and transaxles with internal linkage have no adjustment. Shift cables should not be modified in any way, only replaced or adjusted.

Task C.3 Inspect and replace transaxle gaskets, sealants, seals, and fasteners; inspect sealing surfaces.

A worn outer driveaxle joint may cause a clicking noise while cornering at low speed, but this defect would not cause repeated driveaxle seal failure. A plugged transaxle vent may cause excessive transaxle pressure and repeated driveaxle seal failure.

Transaxle cases have machined mating surfaces that have a very smooth flat finish on them. Not all require a gasket, but they do require some sort of sealant that should be equivalent to the manufacturer's specifications. If a transaxle is assembled without following manufacturer's sealing instructions, leakage from the case will result.

The transaxle mating surfaces should be inspected for warpage with a straightedge before assembly to ensure a proper fit.

Task C.4 Remove and replace transaxle; inspect, replace, and align transaxle mounts.

A misaligned engine and transaxle cradle may cause driveaxle vibrations. Since the lower control arms are connected to this cradle, misalignment of the cradle may cause improper front suspension angles.

When removing a transaxle from a vehicle, it is necessary to install an engine support bracket. This will hold the weight of the engine while the transaxle is being removed. It is installed for the safety of the technician, as well as to avoid damaging the vehicle. You do not have to drain engine oil when removing a transaxle. You should disconnect the negative battery cable. Not all vehicles require you to remove the engine along with the transaxle.

Task C.5 Disassemble and clean transaxle components; reassemble transaxle.

While assembling a manual transaxle it is important to apply gear lube to all of the transaxle parts. Before checking the specifications of the shafts in the transaxle, rotate the shafts to work the gear lube into the bearings. If the gear lube is not worked into the bearings a false measurement may be made.

Before disassembling a transaxle, observe the effort it takes to rotate the input shaft through all forward gears and reverse. Extreme effort in any or all gears may indicate an end play problem or a bent shaft.

Task C.6 Inspect and repair or replace transaxle shift cover and internal shift forks, levers, bushings, shafts, sleeves, detent mechanisms, interlocks, and springs.

Worn dog teeth on the fourth-speed gear would not cause the transaxle to jump out of third gear. A weak detent spring on the 3-4 shift rail may cause the transaxle to jump out of third gear. By not having enough spring pressure, the weak spring could cause this to happen.

The shift forks are used to shift gears, but they are not connected to the forward gears or reverse gear. Also, the shift forks are not connected to the blocker rings or the countershaft. A blocker ring is used to stop a gear from spinning while the vehicle is in motion for gear synchronization. A countershaft is used to change the rotation of the gears on the main shaft and for different gear ratios. The shift forks are connected to the synchronizer assembly. They move the synchronizer sleeve forward or backward to engage the transaxle in a gear.

Task C.7 Inspect and replace input shaft, bearings, and retainers.

Worn dog teeth on the third-speed gear or blocking ring may cause hard shifting or jumping out of third gear, but this problem would not cause a growling noise while driving in third gear. Worn threads in a third-speed blocking ring may cause hard shifting, but this wear would not cause a growling noise in third gear. Worn, chipped teeth on the third speed gear could result in a growling noise while driving in third gear.

Task C.8 Inspect and replace output shaft, gears, thrust washers, bearings, and retainers or snap rings.

Worn dog teeth on the second-speed gear and blocking ring would not result in a growling noise while driving, or accelerating, in second gear, but this problem could not cause hard shifting in second gear. Worn dog teeth on the second-speed gear and blocking ring may cause the transaxle to jump out of second gear.

Inspect all small parts in the transmission for wear. A service manual may list specifications for the thickness of parts, such as thrust washers. If specifications are not available, inspect each part for signs of wear or breakage. Normally all the snap rings, roller bearings, washers, and spacers are replaced during a transaxle overhaul. Most manufacturers sell a small parts kit that includes all of these parts.

Task C.9 Inspect and replace synchronizer hub, sleeve, keys (inserts), springs, and blocking (synchronizing) rings; measure blocking ring clearance.

Before disassembly, always mark the synchronizer sleeve and hub so that these components can be reassembled in their original locations. Synchronizer hubs are not reversible on the shaft, and synchronizer sleeves are not reversible on their hubs.

A worn blocker ring will cause the transmission to have gear clash. If the blocker ring is worn in the cone area (meaning that all of the sharp ridges are dull or gone), the blocker ring will not work properly. A blocker ring should stop a gear from spinning through the sharp ridges in the cone area before the synchronizer sleeve engages the gear.

If the drive gear is not positioned correctly on the output shaft it will not operate. Stripped driven gear teeth is a common cause of an inoperative speedometer. Driven gear slippage on the bottom of the vehicle speed sensor is a common cause of a bad driven gear. Although not very common, the drive gears may be stripped, causing an inoperative speedometer. A mispositioned drive gear is a very unlikely cause of an inoperative speedometer.

Task C.10 Inspect and replace reverse idler gear, shaft, bearings, thrust washers, and retainers or snap rings.

Inspect the reverse idler gear teeth for chips, pits, and cracks. Although worn reverse idler teeth may cause a growling noise while driving in reverse, this problem would not cause a failure to shift into reverse. A broken reverse shifter fork may not allow the transaxle to shift into reverse without causing noise.

The needle bearings on a reverse idler gear should be smooth and shiny. Carefully inspect the needle bearings for wear, burrs, and other problems. Worn or damaged needle bearings should be replaced, or damage to other components may occur.

Task C.11 Inspect, repair, and replace transaxle case mating surfaces, bores, bushings, and vents.

Transaxle case replacement is often required if the case is cracked. Some vehicle manufacturers recommend that the case be repaired, depending on how extensive the damage is. The transaxle case may be repaired with an epoxy-based sealer for some transaxle cracks. Loctite® is not recommended for repairing any transaxle case.

If a threaded area in an aluminum housing is damaged, service kits can be used to insert new threads in the bore. Some threads should never be repaired; check the service manual to identify which ones can be repaired.

Task C.12 Inspect and replace speedometer drive gear, driven gear, and retainers.

A speedometer gear is normally mounted on the output shaft. The output shaft spins at driveline speed. An output shaft will never have a drive gear machined into the output shaft. A drive gear is made out of a plastic nylon-type gear, and the teeth on the drive and driven gear have a helical-type cut on them. The helical-type cut and the plastic-type gear are used for quiet operation.

Whenever a speedometer drive gear is replaced, the driven gear also should be replaced. If a speedometer cable assembly core is damaged, it may be replaced with a new core. The new core must be cut to the same size as the one being replaced and be properly lubricated before installation into the cable housing.

Task C.13 Inspect, test, and replace transaxle sensors and switches.

Most vehicle speed sensors (VSS) are magnetic pickup coil signal generators. Some from the early 1980s, however, are rotary magnetic switches or optical sensors. You can test a magnetic pickup coil with an ohmmeter for coil resistance and for an open or short circuit. Ohmmeter tests are not valid for a magnetic switch or optical VSS.

The signal from a magnetic pickup VSS is an analog sine wave that varies in frequency and amplitude with vehicle speed. The signal from a magnetic switch or an optical VSS is a digital square wave that varies in frequency only. The signal integrity and waveform of any kind of VSS is best tested with an oscilloscope.

Task C.14 Diagnose differential assembly noise and vibration problems; determine needed repairs.

Damaged ring gear teeth would cause a clicking noise while the vehicle is in motion. This problem would not cause differential chatter. Improper preload on differential components, such as side bearings, may cause differential chatter.

If there were a constant whining noise coming from the differential, the noise could not be coming from the side gears or the spider gears. These gears are only used when the vehicle is turning, so if they were damaged, the noise would only be heard on turns. The wrong differential lube could cause damage to the differential parts, but will not cause a whining noise. If the preload and backlash are not set properly, the gear mesh could be too tight and cause a whining noise.

Task C.15 Remove and replace differential assembly.

There are no threaded adjusters on the differential case to adjust bearing preload. The case is not machined exactly to set bearing preload automatically when the two halves are put together.

Engine removal is not always required when removing the transaxle. The transaxle must be removed in order to remove the differential assembly. The entire transaxle does not need to be disassembled to remove the differential assembly.

Task C.16 Inspect, measure, and adjust and replace differential pinion gears (spiders), shaft, side gears, thrust washers, and case.

The side gear end play must be measured individually on each side gear with the thrust washers removed. Side gears with the specified thrust washer have slight end play, but no preload.

The spider gears ride on the pinion shaft, and the bore in the gears should be smooth, shiny, and have no signs of pits or scuffing. The pinion shaft should also be free of pitting and scoring. There is no needle bearing in any of the spider gears.

Task C.17

Inspect and replace differential side bearings.

A side bearing preload adjustment shim is positioned behind one of the side bearing cups. When the differential side bearings are being replaced, they do not need to be packed with grease. The bearings should be lubricated with the differential lube. The differential case does not need to be replaced when the bearings are replaced. The only time a case is replaced is when it shows signs of damage. When new bearings are installed they should be installed using a hydraulic press, and the bearing races should be replaced also. The new bearing will not wear in properly if the old race is used.

Most late-model vehicles use an automatic transmission fluid or a motor oil of 5W30 or 10W30 in manual transaxles. Late-model vehicles require a thinner, more free-flowing fluid because tolerances are a lot tighter than they used to be. A thicker fluid may not provide enough protection with such close or tight tolerances. Also, late-model vehicles need to meet certain gas mileage levels without giving up performance. Thinner transmission fluid will allow for less load on the engine, helping gas mileage and performance.

Task C.18

Measure shaft end play or preload (shim or spacer selection procedure).

While measuring the differential side play to determine the required side bearing shim thickness, a new bearing cup is installed in the transaxle case without the shim. The proper shim thickness is equal to the differential end play plus a specified thickness for bearing preload. While measuring the differential end play, the transaxle case bolts must be tightened to the specified torque. A medium load should be applied to the differential in the upward direction while measuring the differential end play.

When the differential turning torque is less than specified, the shim thickness must be increased behind the side bearing cup in the bell housing side of the transaxle.

Task C.19

Inspect lubrication devices; check fluid level and refill with proper fluid.

Wrong transaxle lubricant may cause burned output shaft bearings. A broken oil feeder behind the front output shaft bearing results in improper output shaft bearing lubrication and burned bearings. If a bearing is not lubricated properly, heat will build and destroy the bearing and possibly other parts of the transaxle.

During transmission or transaxle overhaul, drain and inspect the fluid. Gold-color particles in the fluid are from the wearing of the brass blocking rings on the synchronizers. Metal shavings in the fluid are produced from the wearing of the gears. An excessive amount of shavings in the fluid indicates severe gear and synchronizer wear.

On a cold start, if the fluid is too thick, the vehicle may be hard to shift. This will also cause the transaxle to get poor lubrication on a cold start.

D. Drive (Half) Shaft and Universal Joint/Constant-Velocity (CV) Joint Diagnosis and Repair (Front- and Rear-Wheel Drive) (6 Questions)

Task D.1

Diagnose shaft and universal/CV joint noise and vibration problems; determine needed repairs.

A worn universal joint (U-joint) may cause a squeaking noise that increases in relation to vehicle speed. If the centering ball and socket is worn in a double Cardan U-joint, a heavy vibration may occur during acceleration.

A torsional damper will not cause a clicking noise if it is worn out or bad. If the torsional damper was bad, a shudder would be felt in the vehicle. A constant-velocity (CV) inner joint is not affected when the vehicle is turning. It is only affected when the vehicle suspension is jounced or rebounded while driving. Axle shafts are serviceable; therefore the whole axle shaft does not need to be replaced. An outer CV joint will make a clicking noise when the vehicle is turning. This means the joint ball bearings are bad and the grease is contaminated. The CV joint should be replaced.

Task D.2 Inspect, service, and replace shafts, yokes, boots, and universal/CV joints.

A worn front wheel bearing usually results in a growling noise while cornering or driving straight ahead. A clunking noise while decelerating may be caused by a worn inner driveaxle joint.

A universal joint has prelube on the inside, but this prelube is not sufficient to lubricate the joint when it has been installed in a vehicle. You should grease a new universal joint when it is installed. When a universal joint is greased, you should not pump so much grease into the joint that it squirts out of the caps. When this happens, it damages the seals around the caps and shortens the life of the joint.

Task D.3 Inspect, service, and replace shaft center support bearings.

A worn driveshaft center bearing or an outer rear axle bearing causes a growling noise that is not influenced by acceleration and deceleration.

A center support bearing is not maintenance-free, and it should be maintained and greased as a universal joint. A center support bearing is not part of the driveshaft, but is part of the driveline. A center support bearing is found mainly on trucks and vans and is used to shorten the length of long driveshafts and to decrease pinion angles.

Task D.4 Check and correct propeller shaft balance.

Before checking driveshaft balance, always inspect the shaft for damage. A missing balance weight, accumulation of dirt, or excessive undercoating will affect the balance.

To check driveshaft balance, chalk mark the driveshaft at four locations 90 degrees apart slightly in front of the driveshaft balance weight. Hold a strobe light against the rear axle housing just behind the pinion yoke. Run the vehicle in gear until the driveshaft vibration is the most severe. Point the strobe light at the chalk marks on the driveshaft and note the position of one reference mark. The number that appeared on the strobe light should have two screw-type clamps installed on the shaft near the rear with the heads of the clamp opposite the number that appeared. The vehicle's suspension should have the weight of the vehicle on it when this procedure is performed so that the suspension is at its normal ride height and there are no abnormal pinion angles.

Task D.5 Measure shaft runout.

While measuring driveshaft runout, the dial indicator should be positioned near the center of the driveshaft. The driveshaft should be replaced if the runout is excessive. A driveshaft that is bent or damaged in any way should be replaced; repairs to a damaged driveshaft should not be attempted.

To check driveshaft runout, the vehicle should be raised and a dial indicator with a magnetic base should be installed under the vehicle near the center of the driveshaft. The surface of the shaft should be wiped off or cleaned in case it is rusted or has dirt buildup that may affect the reading on the dial indicator. Rolling a driveshaft on a flat surface is not an accurate way of checking the driveshaft runout.

Task D.6 Measure and adjust shaft angles.

An out-of-balance driveshaft will cause a vibration that increases as the speed of the vehicle increases. Out-of-balance wheels will also cause the vehicle to vibrate. The vibration intensity will increase as the vehicle accelerates.

An excessive driveshaft angle can cause the universal joint to bind. If the joint is on an extreme angle, it will not be able to rotate properly. Excessive angles will not cause a loud humming noise; this noise is usually due to gears or abnormal wear on tires. Excessive angles will not cause the transmission mount to be damaged, nor will they cause the pinion gear in the axle to be damaged.

E. Rear-Wheel DriveAxle Diagnosis and Repair (7 Questions)

1. Ring and Pinion Gears (3 Questions)

Task E.1.1 Diagnose noise, vibration, and fluid leakage problems; determine needed repairs.

Since the side gears are only turning while cornering, they do not cause a whining noise while driving straight ahead. Improper ring gear and pinion adjustments may cause this kind of noise.

If the differential fluid is too full, excessive pressure may build up and cause the differential fluid to leak past a seal. If the vent becomes plugged—this is common with vehicles that are subjected to extreme weather conditions—it will cause excessive pressure in the differential housing and a leak will occur. Axle shaft bearings that are worn will cause the axle shaft to apply load on the axle shaft seal, and the seal will fail. Light-viscosity lubricant should not cause a leak, but it may lead to abnormal or premature wear of internal moving parts.

Task E.1.2 Inspect and replace companion flange and pinion seal; measure companion flange runout.

A loose pinion nut allows pinion shaft end play, resulting in a clunking noise on acceleration and deceleration. Insufficient pinion nut torque will affect the pinion bearing. Preload will not cause a growling noise.

To check pinion flange runout, remove the driveshaft and mount a dial indicator against the face of the flange. Rotate the flange and note the readings on the dial indicator; these are the runout readings. If the flange is removed, you cannot measure pinion flange runout.

Task E.1.3 Measure ring gear runout; determine needed repairs.

Excessive side bearing preload, side gear end play, or ring gear bolt torque will not cause excessive ring gear runout.

A hunting-type pinion gear and ring gearset do not require the gearset to be timed. Many ring gearsets and pinion gearsets have timing marks that must be aligned when assembling the differential.

Task E.1.4 Inspect and replace ring and pinion gear set, collapsible spacers, sleeves (shims), and bearings.

The collapsible spacer and the antilock brake system (ABS) toothed ring (if equipped) should both be replaced when the pinion shaft has been disassembled. The ABS ring will be damaged when it is removed. The collapsible spacer will not set the pinion preload correctly if it is not replaced.

Task E.1.5 Measure and adjust drive pinion depth.

The base pinion depth setting is the distance between the nose of the pinion gear and the center of the axles or differential case bearing bores. The pinion gear depth is normally adjusted by installing shims onto the pinion mounting; these shims change the depth of pinion and ring gear mesh. The pinion gear may be marked with additional adjustment in addition to the base pinion depth setting. If the pinion is marked with a plus or minus number (e.g. +2 or –4), add or subtract that many thousandths from the base setting. A pinion that has no markings should be installed to the base setting only.

The pinion nut must never be loosened to obtain the specified turning torque. The pinion bearings should be lubricated when the turning torque is measured.

Task E.1.6 Measure and adjust drive pinion bearing preload (collapsible spacer or shim type).

The left bearing adjuster nut must be loosened and the right bearing adjusting nut tightened to increase ring gear backlash.

With either excessive ring gear toe contact or low flank contact, the pinion gear must be moved away from the ring gear.

Worn pinion bearings may cause a growling noise while driving straight ahead. The side gears and the differential case may cause a vibration while cornering.

Task E.1.7 Measure and adjust differential (side) bearing preload and ring and pinion backlash (threaded cup or shim type).

Ring gear runout and case side play should be measured before removing the ring gear and case assembly. The side bearing caps should be marked in relation to the case before removal. The side bearings should be lubricated before installation.

The side bearings must be in good condition before measuring case runout. The ring gear runout should be measured before the case runout.

Task E.1.8 Perform ring and pinion tooth contact pattern checks; determine needed adjustments.

In a limited slip differential, the steel plates are retained in the case; each clutch set contains a preload spring, and a special lubricant is required. The friction plates are splined to the clutch hub.

The wrong lubricant or worn friction and steel plates in a limited slip differential may cause chattering while turning a corner.

2. Differential Case Assembly (2 Questions)

Task E.2.1 Diagnose differential assembly noise and vibration problems; determine needed repairs.

Using clay between the ring gear and pinion gear teeth is not an accurate way of checking the tooth contact pattern. Powder chalk and paint are sometimes used to check the tooth contact pattern on gear setup. A used gearset will have a shiny pattern on the gear teeth that can be visually inspected.

Excessive ring gear backlash could cause a clunk noise on deceleration. A universal joint could also cause this problem, but will also probably have a vibration along with the noise.

Task E.2.2 Remove and replace differential assembly.

On some types of axles, the pinion gear and differential assembly all come out of the axle housing as one assembly. The bearing caps should always be marked when removed to ensure they go back together correctly. The axle shafts must be removed before the differential assembly will come out. The shim packs and bearing races should be kept in order.

To get a rotating torque measurement on the pinion gear, the differential case must be removed. If the load of the case, ring gear, and the axle shafts are included, you measure the rotating torque of the axle assembly. Use an inch-pound torque wrench with a needle-type indicator to get an accurate reading.

Task E.2.3 Inspect, measure, adjust, and replace differential pinion gears (spiders), shaft, side gears, thrust washers, and case.

Use feeler gauges to check the clearance between differential side gears and thrust washers. Remove the differential side gears and pinions. Inspect all gears for abnormal wear, heat damage, chipped teeth, brinelling, and other wear. Inspect the differential pinion shaft for wear, cracks, lack of lubrication, and heat damage. When assembling differential gears and side gears, place the side gears and washers into the case. Then walk the pinions around the side gears until they align with the shaft hole; insert the shaft, spacer block, and lockpin. After installation, rotate the gears a few times and recheck thrust washer clearance with feeler gauges.

Task E.2.4 Inspect and replace differential side bearings.

Inspect the side bearings for wear, particularly signs of the inner races turning on the bearings. Replace worn parts. Inspect thrust washer surfaces for nicks and abnormal wear. If new side bearings are required, lubricate the surfaces that contact the case. Install shims if required and drive or press the bearings onto the case. Apply force on the inner cone, not on the rollers. If new bearings are installed, install new outer races (cups) in the axle housing or differential carrier.

Task E.2.5 Measure differential case runout; determine needed repairs.

To check differential case runout, place the case in a set of V-blocks. Then place a dial indicator against the ring gear mounting flange and rotate the case to measure axial runout. Move the dial indicator pointer to the ring gear hub area of the case and rotate the case again to measure radial runout. If runout is out of limits in either direction, replace the case.

3. Limited Slip Differential (1 Question)

Task E.3.1 Diagnose limited slip differential noise, slippage, and chatter problems; determine needed repairs.

Weak spring tension, a broken spring, and worn friction plates are common symptoms of a failed limited slip differential. As the friction material wears off the friction plates, they will not grip as well. Also, worn plates are thinner and allow the tension spring to stretch out more, reducing tension. Adding the wrong differential fluid to the limited slip differential may cause failure. The friction plates are not under excessive load, and therefore the teeth on the plates are not likely to strip out.

There are no special copper plates in a limited slip differential. The plates are made of steel, and they cannot absorb lubricant. The friction modifier is not used to help the plates grab better, it is added to eliminate differential chatter while turning corners. Since the differential is not in motion on a highway, the friction modifier would not be used for this purpose.

Task E.3.2 Inspect, flush, and refill with correct lubricant.

When filling a transmission or differential case, fluid will come out of the fill hole as it reaches the full level. After draining differential fluid, inspect it for excessive metal particles. Silver- or steel-colored particles are signs of gear or bearing wear. Copper- or bronze-colored particles are signs of limited slip clutch disc wear.

Task E.3.3 Inspect, adjust, and replace clutch (cone/plate) pack.

The friction plates have a minimum thickness specification and should be measured with a micrometer. The friction plate has to be removed from the clutch pack for this measurement. There is no way to measure a friction plate with a feeler gauge.

4. Axle Shafts (1 Question)

Task E.4.1 Diagnose rear axle shaft noise, vibration, and fluid leakage problems; determine needed repairs.

Excessive axle shaft runout may cause a vibration at 60 mph (97 km/h), but this problem would not cause a vibration when accelerating at lower speeds.

Task E.4.2 Inspect and replace rear axle shaft wheel studs.

When replacing a wheel stud, a hammer can be used to carefully tap the old stud out, but when installing a wheel stud, an installation tool should be used to avoid damage to the new stud. A torch should never be used to burn out a wheel stud because damage to the axle shaft and axle seal may occur.

Task E.4.3 Remove, inspect, and replace rear axle shafts, seals, bearings, and retainers.

A worn axle shaft bearing can cause axle shaft seal failure. When the bearings wear out, axle side movement applies a greater load on the seal lip. A plugged axle vent can cause pressure buildup and force fluid past the seal. Scored axle shafts in the seal area will damage the seal lip and cause the seal to fail. Heat from the brakes will not cause the seal to melt because it is made of a heat-treated rubber.

The axle shaft seal comes with a sprayed on sealant and does not require any other sealant. The sealing lip of the axle shaft seal should be lubricated with a light coating of gear lube to prolong the seal life.

If the axle shaft is damaged near the seal area, the shaft should be discarded and replaced with a new one. The axle shaft should be stood up straight when removed to avoid damage to the axle splines and the wheel studs. The seals should be replaced every time an axle shaft is removed because they can be damaged by the axle splines. Some axle shafts require a slide hammer to remove them.

Task E.4.4 Measure rear axle flange runout and shaft end play; determine needed repairs.

To measure the axle shaft end play, you will need to remove the wheel and tire assembly and the brake drum. A dial indicator is mounted or clamped to the axle housing or suspension. The axle shaft must be pushed in to the housing all the way until it stops. Rest the dial indicator head on the face of the axle shaft flange. With the dial indicator set to zero, pull out on the axle shaft. The resulting dial indicator reading is the axle shaft end play. The vehicle differential cover does not need to be removed.

Excessive runout could be caused by a bent axle shaft. A worn C-lock will cause excessive end play. A worn bearing will cause fluid leakage. A bent housing is considered major damage and is noticeable.

F. Four-Wheel Drive Components Diagnosis and Repair (7 Questions)

Task F.1 Diagnose four-wheel drive assembly noise, vibration, shifting, and steering problems; determine needed repairs.

Worn U-joints may cause a squeaking or clunking noise, and a vibration while driving straight ahead. Worn outer front driveaxle joints on a 4WD vehicle may cause a vibration while cornering.

When a vacuum-operated 4WD does not shift into 4WD, the engine vacuum may be low or the vacuum motor at the front differential may be damaged. Another cause could be bad or disconnected vacuum lines.

Task F.2 Inspect, adjust, and repair transfer case manual shifting mechanisms, bushings, mounts, levers, and brackets.

Specification measurements should be taken and recorded to aid in the installation of parts that have tolerances. All parts should be cleaned and lubricated before the assembly of the part. All components should be inspected for wear or damage.

The annulus gear is locked to the case so it cannot rotate. In four-wheel drive (4WD) low, the transfer case input shaft is driving the sun gear, which, in turn, is driving the planetary carrier.

Task F.3 Remove and replace transfer case.

Transmission and transfer cases are removed as an assembly. Most newer 4WD vehicles have the transfer case bolted to the rear of the transmission. Not all transfer cases are made of cast iron; most new models have a lightweight aluminum case.

Task F.4 Disassemble and clean transfer case and components; reassemble transfer case.

A plugged transfer case vent may cause seal leakage. A remote transfer case vent helps keep water out of the transfer case.

Task F.5 Inspect and service transfer case and internal components; check lube level.

If the drive chain in the transfer case is stretched, and the drive sprocket teeth are worn, the chain could slip on the teeth and cause a loud clicking noise under acceleration.

If the four-wheel drive (4WD) on the vehicle is not used often, the shift linkage could become rusted, and it will not work easily or at all. The shift linkage should be inspected for bushings that may be worn or deteriorated and need replacement. If the shift linkage does not move the linkage its full range, the transfer case may not operate in 4WD. The transfer case would still shift into gear if the front driveshaft universal joints are bound. Low fluid may cause damage to internal components, but the transfer case will still engage. A manual shift transfer case has no electronic shift motor.

Task F.6 Inspect, service, and replace front-drive (propeller) shafts and universal/CV joints.

Worn universal joints (U-joints), front axle drive joints, and incorrect driveshaft angles may cause a vibration that is more noticeable when changing throttle position.

A snap ring holds the inner tripod joint on the axle shaft, and a special swaging tool may be necessary to tighten the outer boot clamp. A worn outer constant-velocity (CV) joint may cause a clicking noise while cornering. All of the grease supplied with the joint should be used; this is the correct amount to be used.

Task F.7 Inspect, service, and replace front-driveaxle knuckles and driving shafts.

It is not necessary to remove the axle assembly to remove the driveaxles. A slide hammer may be needed to remove the driveaxles from the housing.

Task F.8 Inspect, service, and replace front wheel bearings and locking hubs.

Neither the automatic locking hubs nor the caps should be packed with grease. If they are packed with grease, they will not operate properly; the parts must move freely.

If only one bearing is bad, replace only that bearing. When a wheel bearing is replaced, the bearing race should also be replaced. The new bearing may fail if the race is not replaced.

Task F.9 Check transfer case and front axle seals and remote vents.

The only purpose of a remote vent is to keep moisture out of a differential assembly. This is needed on a four-wheel drive (4WD) vehicle in case the axle or transfer case is submerged in water.

Task F.10 Diagnose, test, adjust, and replace electrical/electronic components of 4WD systems.

An electronic-shift transfer case has all of the shift linkage inside the case. If an electronic-shift transfer case has a problem with shifting, an electrical component would be a likely cause of failure.

Sample Test for Practice

Sample Test

Please note the letter and number in parentheses following each question. They match the overview in section 4 that discusses the relevant subject matter. You may want to refer to the overview using this cross-referencing key to help with questions posing problems for you.

1. Repeated extension housing seal failure may be caused by:
   A. a scored driveshaft yoke.
   B. excessive output shaft end play.
   C. excessive input shaft end play.
   D. a worn output shaft bearing. (B.3)

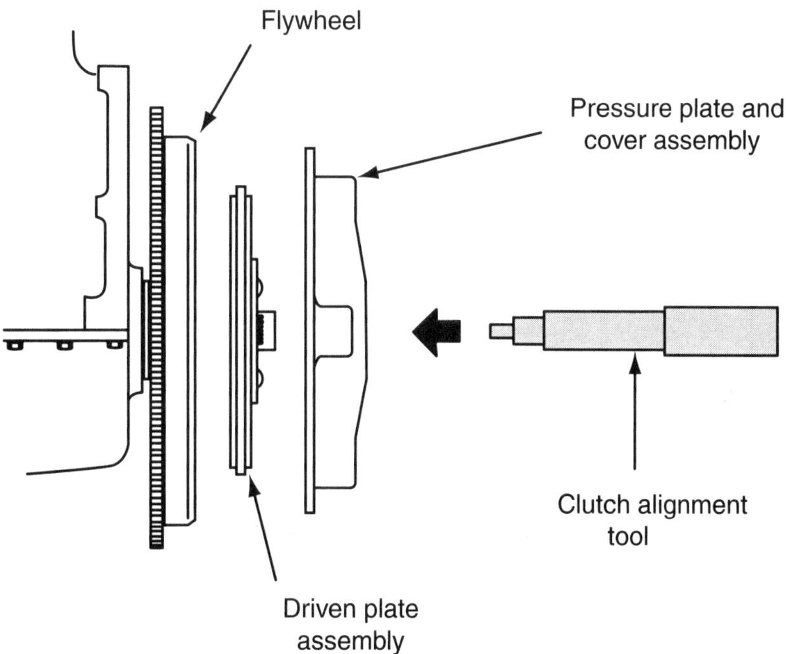

2. During manual transmission removal and replacement:
   A. the driveshaft may be installed in any position on the differential pinion gear flange.
   B. the transmission weight may be supported by the input shaft in the clutch disc hub.
   C. the engine support fixture should be installed after the transmission-to-engine bolts are loosened.
   D. the clutch disc must be aligned with an aligning tool before transmission installation. (B.4)

3. When replacing a wheel stud, Technician A says that you should use a hammer to remove and install the axle shaft flange. Technician B says you will need to heat the wheel studs with a torch to remove them. Who is right?
   A. A only
   B. B only
   C. Both A and B
   D. Neither A nor B (E.4.2)
4. Technician A says the collapsible pinion shaft spacer may be reused if the differential is disassembled and overhauled. Technician B says that the antilock brake system (ABS) toothed ring on the pinion shaft may be reinstalled if it is removed from the shaft. Who is right?
   A. A only
   B. B only
   C. Both A and B
   D. Neither A nor B (E.1.4)
5. Technician A says the dial indicator should be positioned near the front of the driveshaft to measure driveshaft runout. Technician B says if the driveshaft runout is excessive, the driveshaft may be straightened in a hydraulic press. Who is right?
   A. A only
   B. B only
   C. Both A and B
   D. Neither A nor B (D.5)

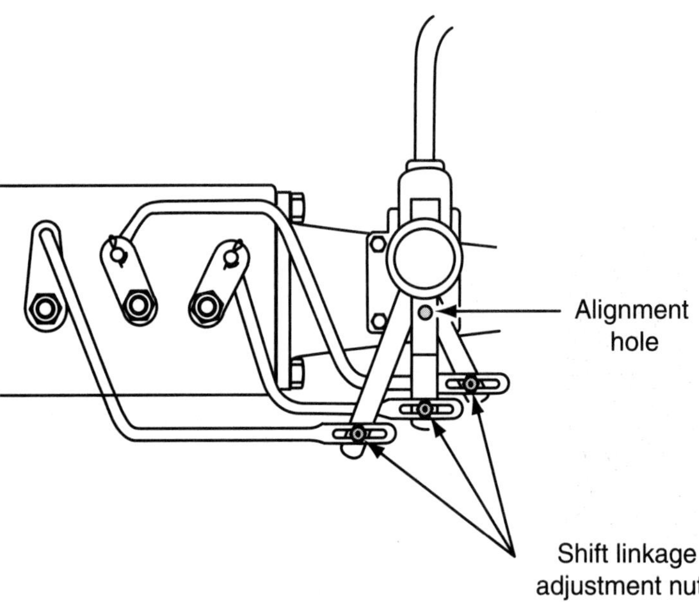

6. The shift lever adjustment is usually performed with the transmission in:
   A. neutral.
   B. first gear.
   C. second gear.
   D. reverse gear. (B.2)

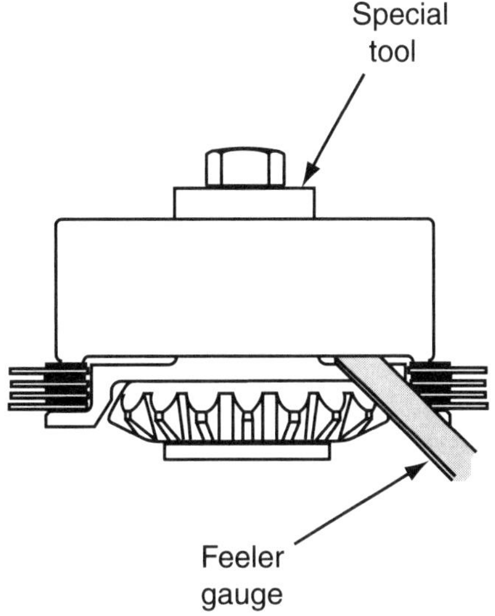

7. The measurement in the figure shown determines the proper:
   A. friction plate thickness.
   B. steel plate thickness.
   C. shim thickness.
   D. preload spring tension. (E.1.8)

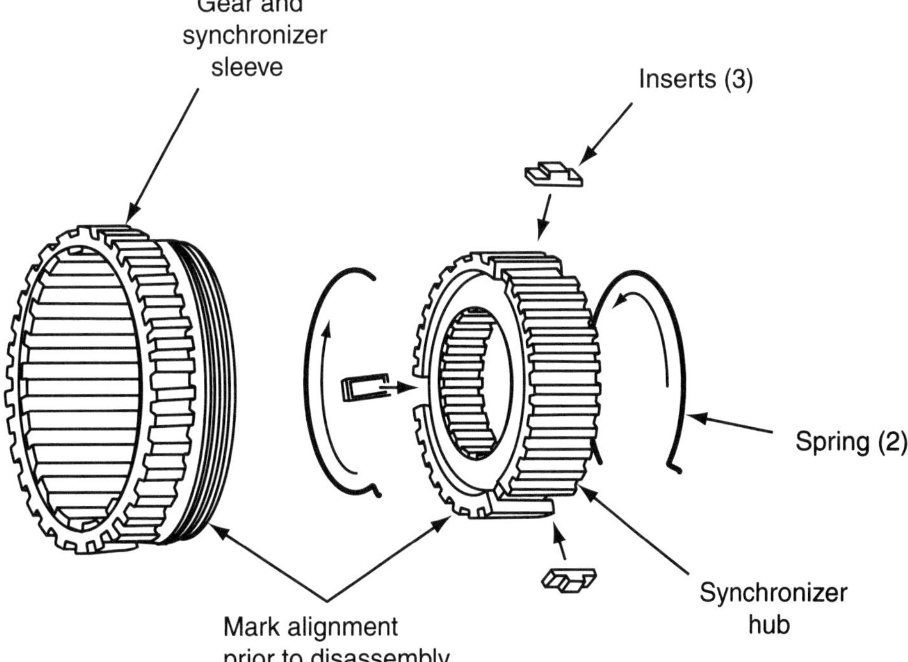

8. While inspecting synchronizer assemblies on a manual transmission:
   A. the dog teeth tips on the blocking rings should be flat with smooth surfaces.
   B. the threads in the cone area of the blocker rings should be sharp and not dull.
   C. the clearance is not important between the blocking ring and the matching gear's dog teeth.
   D. the sleeve should fit snugly on the hub and offer a certain amount of resistance to movement. (B.9)

9. Technician A says that if the synchronizer sleeve does not slide smoothly over the blocker ring and gear teeth, shifting may not occur. Technician B says the synchronizer hub and sleeve must be marked in relation to each other before disassembly. Who is right?
   A. A only
   B. B only
   C. Both A and B
   D. Neither A nor B (B.5)

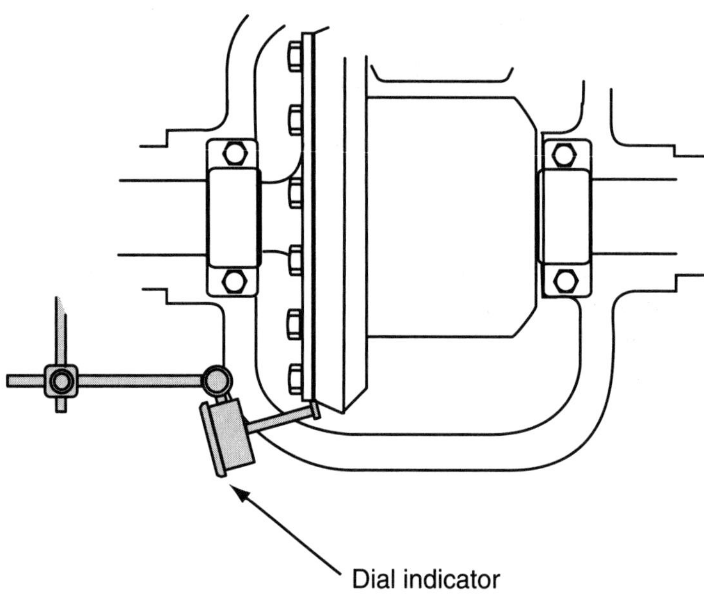

10. Excessive ring gear runout on the dial indicator shown in the figure may be caused by excessive:
    A. differential case runout.
    B. side bearing preload.
    C. side gear end play.
    D. ring gear bolt torque. (E.1.3)
11. A front-wheel drive car has a clunking noise while decelerating. Technician A says this noise may be caused by a worn inner drive axle joint. Technician B says this noise may be caused by a front wheel bearing. Who is right?
    A. A only
    B. B only
    C. Both A and B
    D. Neither A nor B (D.2)

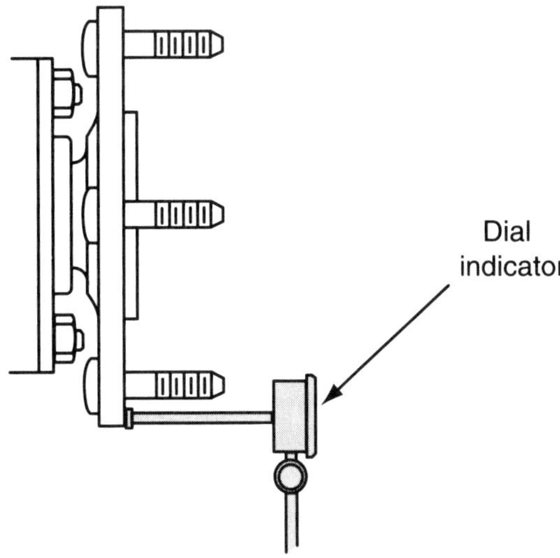

12. The cause of excessive runout on the dial indicator shown in the figure could be:
    A. incorrect differential bearing preload.
    B. a bent axle shaft.
    C. incorrect pinion depth.
    D. worn limited slip clutch discs. (E.4.4)

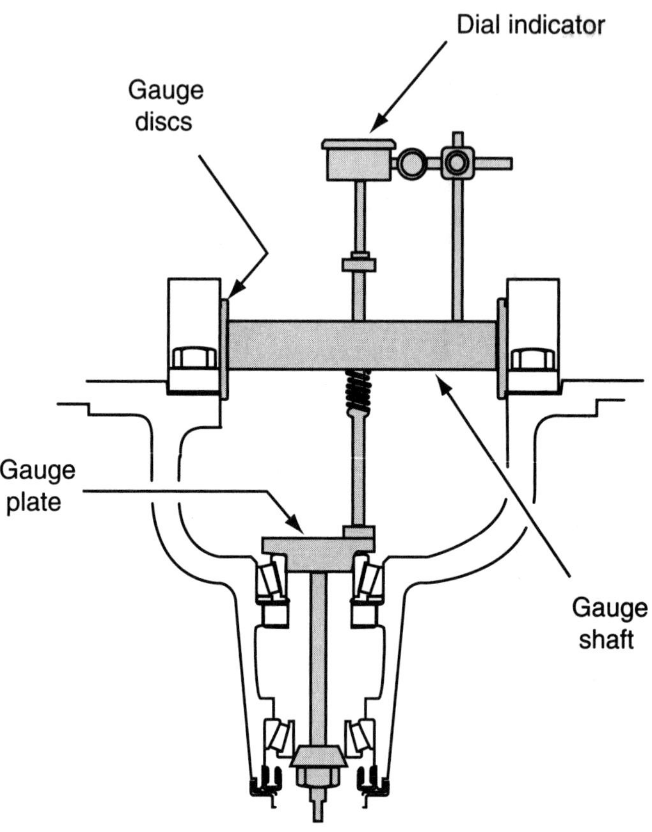

13. In the figure shown, a 0.063 inch (1.6 mm) shim fits between the gauge block and the gauge tube with a light drag, and the pinion gear is marked +3. The proper pinion shim depth is:
    A. 0.060 inch (1.52 mm).
    B. 0.041 inch (1.04 mm).
    C. 0.066 inch (1.68 mm).
    D. 0.069 inch (1.75 mm). (E.1.4)
14. Technician A says sagged transmission mounts may cause improper driveshaft angles on a rear-wheel drive car. Technician B says improper driveshaft angles may cause a constant-speed vibration when the vehicle is accelerated and decelerated. Who is right?
    A. A only
    B. B only
    C. Both A and B
    D. Neither A nor B (A.11)
15. In a manual transmission, the reverse idler gear is in mesh with what gear only?
    A. First gear
    B. Fifth gear
    C. Third and fourth gear
    D. Reverse gear (B.11)

16. While discussing a four-wheel drive vehicle with a vibration problem that is more noticeable while cornering, Technician A says the U-joints may be worn. Technician B says the outboard front axle joints may be worn. Who is right?
    A. A only
    B. B only
    C. Both A and B
    D. Neither A nor B (F.1)
17. The countergear shaft and needle bearings are pitted and scored. Technician A says the transmission may have a growling noise with the engine idling, with the transmission in neutral, and with the clutch pedal released. Technician B says the transmission may have a growling noise while driving in any gear. Who is right?
    A. A only
    B. B only
    C. Both A and B
    D. Neither A nor B (B.1)

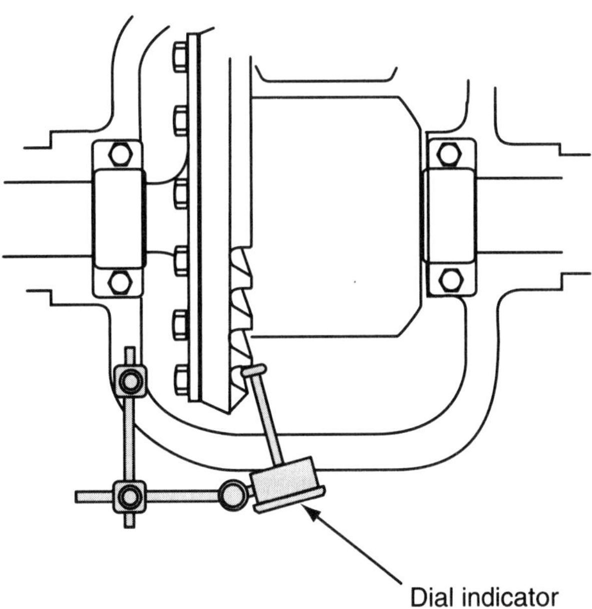

18. In the figure shown, the ring gear backlash and side play are zero. Right and left-side bearing adjustment nuts are determined while facing the differential from the rear. To obtain the proper ring gear backlash:
    A. tighten the right- and left-side bearing adjusters.
    B. loosen the left-side bearing adjuster.
    C. loosen the left-side bearing adjuster and tighten the right-side bearing adjuster.
    D. loosen the right-side bearing adjuster and tighten the left-side bearing adjuster. (E.1.6)
19. A transmission has a growling noise in first gear only. The cause of the problem could be:
    A. a worn first-gear blocker ring.
    B. a worn first-gear synchronizer ring.
    C. a chipped or worn first-speed gear tooth.
    D. a worn, rough main shaft bearing. (B.8)

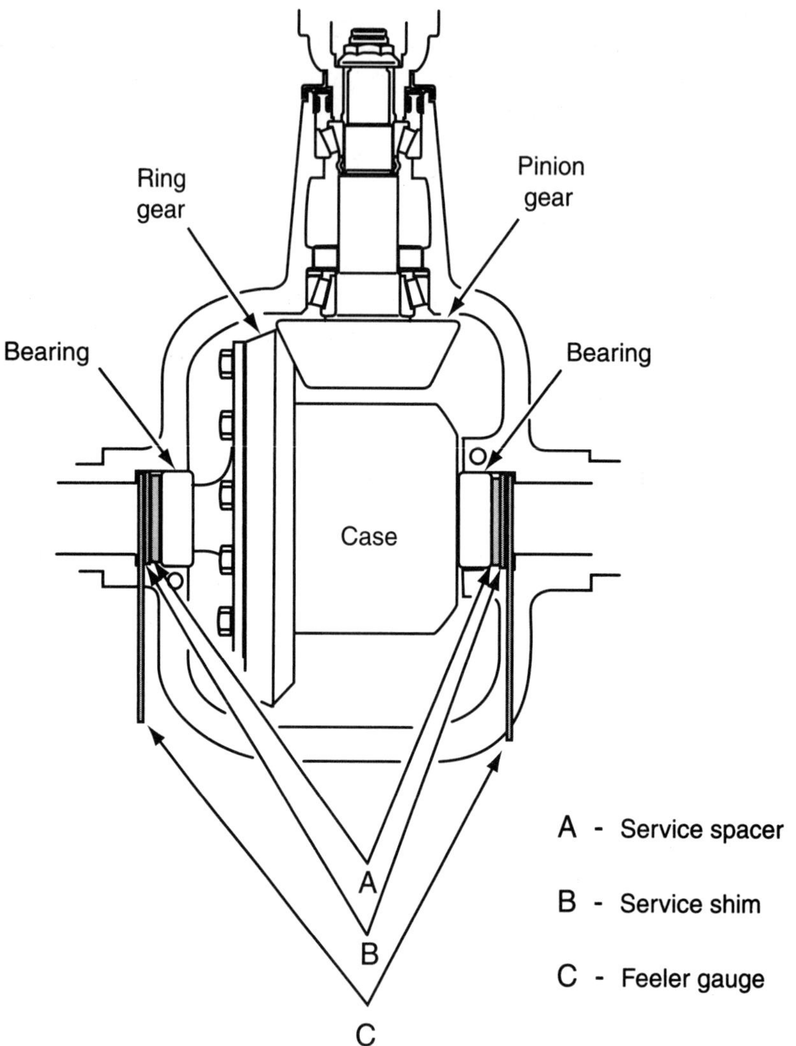

EXAMPLE			
.250"	Ring gear side Combined total of: Service spacer (A) Service shim (B) Feeler gauge (C)	Opposite side Combined total of: Service spacer (A) Service shim (B) Feeler gauge (C)	.265"
-.010" .240"	To maintain proper backlash (.005"–.008") ring gear is moved away from the pinion by subtracting .010" shim from ring gear side and adding .010" to the other side		+.010" .275"
+.004"	To obtain proper preload on side bearings, add .004" shim to each side		+.004"
.244"	Shim dimension required for ring gear side	Shim dimension required for opposite side	.279"

20. Technician A says if the ring tooth contact pattern indicates pinion tooth contact on the toe of the pinion gear, the pinion gear should be moved toward the ring gear. Technician B says if the pinion gear teeth have low flank contact on the ring gear teeth, the pinion gear should be moved toward the ring gear. Who is right?
    A. A only
    B. B only
    C. Both A and B
    D. Neither A nor B (E.1.6)
21. Technician A says a misaligned engine and transaxle cradle may cause drive axle vibrations. Technician B says a misaligned engine and transaxle cradle may cause improper front suspension angles. Who is right?
    A. A only
    B. B only
    C. Both A and B
    D. Neither A nor B (C.4)
22. While discussing repeated transaxle drive axle seal leakage and replacement, Technician A says this problem may be caused by a plugged transaxle vent. Technician B says this problem may be caused by a worn outer drive axle joint. Who is right?
    A. A only
    B. B only
    C. Both A and B
    D. Neither A nor B (C.3)
23. Excessive input shaft end play in the transmission may cause the transmission to jump out of:
    A. first gear.
    B. second gear.
    C. fourth gear.
    D. reverse gear. (B.12)

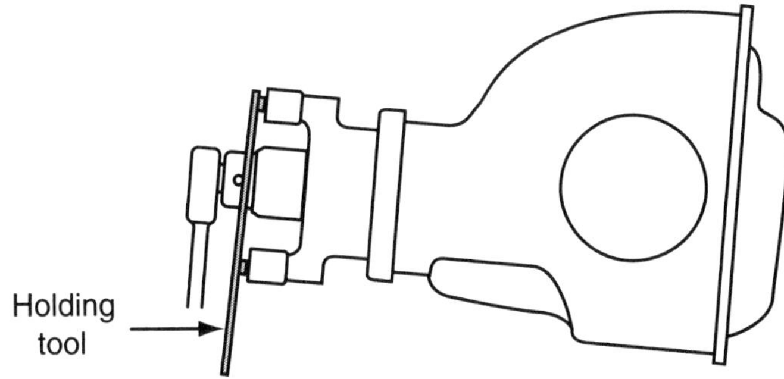

24. Technician A says that insufficient pinion nut torque may cause a clunking noise during acceleration or deceleration. Technician B says that insufficient pinion nut torque may cause a growling noise with the vehicle in motion. Who is right?
    A. A only
    B. B only
    C. Both A and B
    D. Neither A nor B (E.1.2)

25. While a vehicle is moving straight ahead, the differential produces a whining noise. Technician A says that the differential side gears may be damaged. Technician B says that the ring gear and pinion adjustments may be incorrect. Who is right?
    A. A only
    B. B only
    C. Both A and B
    D. Neither A nor B (E.1.1)
26. The needle bearings between the output shaft and the output shaft gears are scored and blue from overheating. Technician A says the transaxle may have been filled with the wrong lubricant. Technician B says the projection is broken off the oil feeder behind the front output shaft bearing. Who is right?
    A. A only
    B. B only
    C. Both A and B
    D. Neither A nor B (C.19)

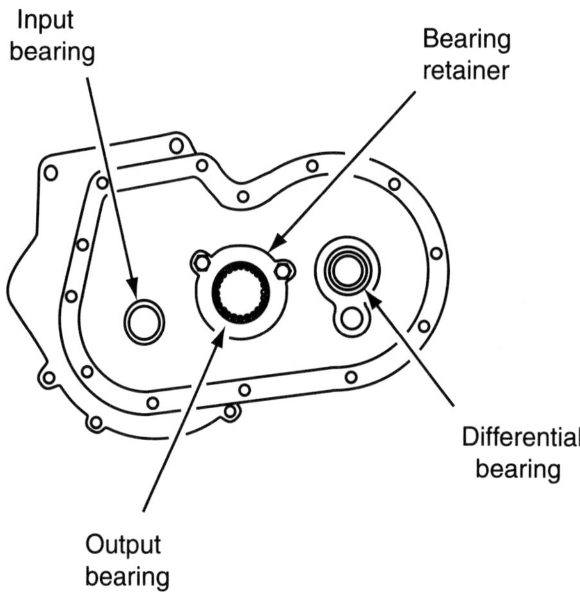

27. While discussing differential side bearing preload in the transaxle case shown in the figure, Technician A says differential bearing preload is adjusted by rotating a threaded adjuster on each side of the differential bearings. Technician B says the differential bearing preload is automatically adjusted when the case halves are reassembled. Who is right?
    A. A only
    B. B only
    C. Both A and B
    D. Neither A nor B (C.15)
28. In some transaxles, the speedometer drive gear is mounted on:
    A. the input shaft.
    B. the transfer gear.
    C. the differential case.
    D. the drive axle inner hub. (C.12)

29. A transaxle shifts normally into all forward gears, but it will not shift into reverse gear; there is no evidence of noise while attempting this shift. Technician A says the reverse shifter fork may be broken. Technician B says the reverse idler gear teeth may be worn. Who is right?
    A. A only
    B. B only
    C. Both A and B
    D. Neither A nor B (C.10)
30. Technician A says synchronizer hubs on a manual transaxle are reversible on the shaft on which they are mounted. Technician B says synchronizer sleeves on a manual transaxle are reversible on their matching hub. Who is right?
    A. A only
    B. B only
    C. Both A and B
    D. Neither A nor B (C.9)
31. A five-speed manual transaxle has a growling and rattling noise in third gear only. The cause of this problem could be:
    A. worn, chipped teeth on the third-speed gear on the input shaft.
    B. worn dog teeth on the third-speed gear on the input shaft.
    C. worn dog teeth on the third-speed synchronizer blocker ring.
    D. worn threads in the cone area of the third-speed blocker ring. (C.7)
32. While discussing the assembly of a manual transaxle, Technician A says that the bearings and gears do not need to be lubricated, they will be lubricated when the transaxle is filled with fluid. Technician B says that the bearing and gears should be lubricated while assembling a manual transaxle. Who is right?
    A. A only
    B. B only
    C. Both A and B
    D. Neither A nor B (C.5)

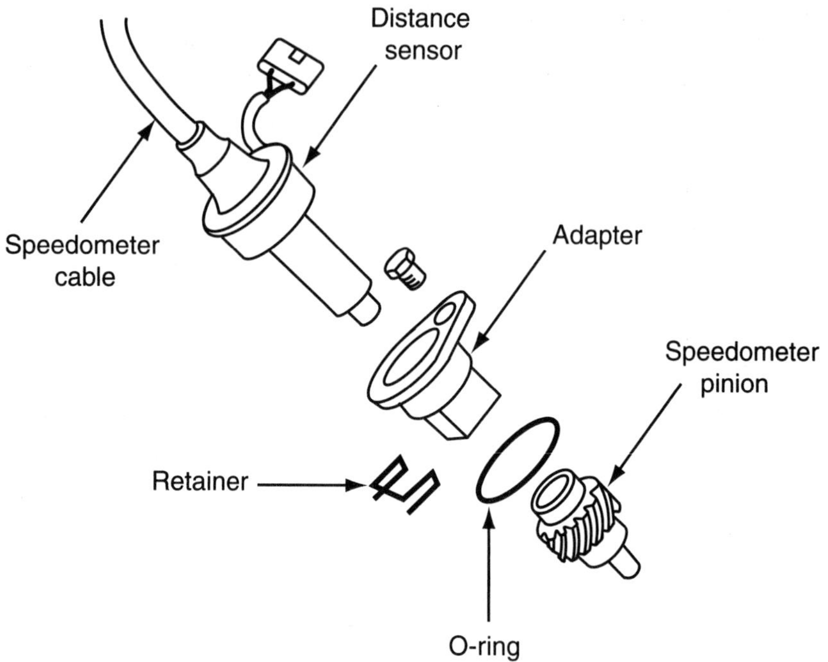

33. Erratic speedometer operation with the speedometer drive shown in the figure may be caused by:
    A. a worn adapter bushing and distance sensor.
    B. a worn extension housing bushing.
    C. excessive transmission main shaft end play.
    D. a distance sensor electrical problem. (B.14)

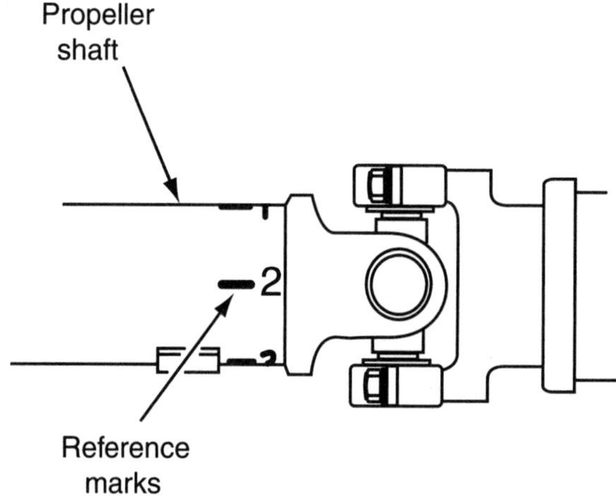

34. The marking shown in the figure is required for:
    A. driveshaft removal and replacement.
    B. universal joint removal and replacement.
    C. driveshaft runout measurement.
    D. driveshaft balance testing. (D.4)

35. While discussing manual transmission lubricants, Technician A says compared to hypoid gear oil, ATF reduces friction and improves fuel economy. Technician B says the thicker the gear oil the lower the viscosity number. Who is right?
    A. A only
    B. B only
    C. Both A and B
    D. Neither A nor B (B.17)

36. Case bearing preload is correct in a shim-type differential, but backlash is excessive. To correct this, you should:
    A. remove shims from the right side and add shims to the left.
    B. add an equal number of shims to both sides.
    C. remove shims from the right side and add shims to the left.
    D. remove an equal number of shims from both sides. (C.17)

37. Technician A says an epoxy-based sealer may be used to repair a crack in some transaxle cases. Technician B says some cracks in transaxle cases may be repaired with Loctite®. Who is right?
    A. A only
    B. B only
    C. Both A and B
    D. Neither A nor B (C.11)

38. All of the following statements about limited slip differentials are true EXCEPT:
    A. friction plates are splined to the sidegears.
    B. steel plates are splined to the rear axle shafts.
    C. each clutch set contains a preload spring.
    D. a special lubricant is required. (E.3.3)

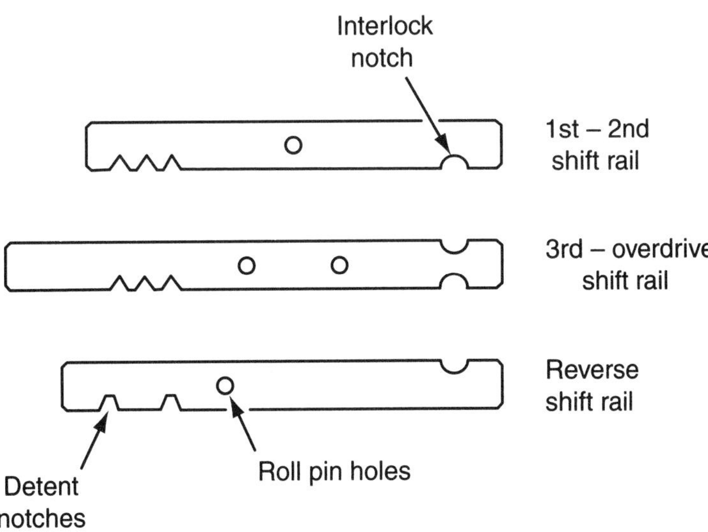

39. A bent shift rail will most likely cause:
    A. the transmission to jump out of gear.
    B. hard shifting in some gears.
    C. gear clash during some shifts.
    D. gear noise in some gears. (B.6)

40. All of the following statements are true about removing and installing axle shafts EXCEPT:
    A. The shaft near the seal area should be inspected and repaired with sandpaper, if needed.
    B. The axle shaft seals should be replaced, not reused.
    C. The axle shaft should be stood straight up when removed.
    D. Some axles require a slide hammer to remove them. (E.4.3)
41. A manual transmission, when in neutral, has a growling noise with the engine idling and the clutch released. The noise disappears when the clutch pedal is depressed. The cause of this noise could be a worn:
    A. pilot bearing in the crankshaft.
    B. input shaft and pilot bearing contact area.
    C. input shaft bearing.
    D. mainshaft ball bearing. (B.7)
42. In a hydraulic clutch, the clutch fails to disengage properly when the clutch pedal is fully depressed. The cause of this problem could be:
    A. less than specified clutch pedal free play.
    B. air in the clutch hydraulic system.
    C. worn clutch facings.
    D. a scored pressure plate. (A.3)
43. The second-speed gear dog teeth and blocking ring teeth are badly worn. This problem may cause:
    A. a growling noise while driving in second gear.
    B. a vibration while accelerating in second gear.
    C. hard shifting in second and third gear.
    D. the transaxle to jump out of second gear. (C.8)
44. A fully synchronized four-speed manual transaxle experiences gear clash in all forward gears and reverse. The cause of this problem could be:
    A. a worn fourth-gear synchronizer.
    B. the clutch disc sticking on the input shaft.
    C. excessive main shaft end play.
    D. a worn 3-4 shifter fork. (C.1)
45. When diagnosing a rear-wheel drive differential that vibrates only while turning a corner, Technician A says the drive pinion bearings may be worn, and the pinion bearing preload is less than specified. Technician B says the bearing surfaces between the side gears and the differential case may be damaged. Who is right?
    A. A only
    B. B only
    C. Both A and B
    D. Neither A nor B (E.1.6)
46. While discussing backup lamp switches, Technician A says that a backup lamp switch is located in the transmission and is normally open. Technician B says that a backup lamp switch can have power to the switch in RUN and is closed when the vehicle is shifted into reverse. Who is right?
    A. A only
    B. B only
    C. Both A and B
    D. Neither A nor B (B.16)

47. All of the following statements about differential case and ring gear removal and replacement are true EXCEPT:
    A. The ring gear runout should be measured before removal of the case and ring gear assembly.
    B. The case side play should be measured before removal of the case and ring gear assembly.
    C. The side bearing caps should be marked in relation to the housing before removal of the case and ring gear assembly.
    D. The side bearing should be clean and dry before installation of the case and ring gear assembly. (E.1.7)
48. Premature wear in the extension housing bushing will most likely be caused by:
    A. worn speedometer drive and driven gears.
    B. metal burrs on the rear transmission mating surfaces.
    C. a plugged transmission vent opening.
    D. excessive transmission main shaft end play. (B.13)
49. A manual transaxle chatters while driving straight ahead. Technician A says the ring gear and pinion gear teeth may be worn and chipped. Technician B says there may be improper preload on the differential components. Who is right?
    A. A only
    B. B only
    C. Both A and B
    D. Neither A nor B (C.14)
50. While discussing a clutch with an adjustable linkage, Technician A says the clutch pedal free play adjustment sets the distance between the release bearing and the pressure plate fingers. Technician B says a worn release bearing is most noisy when the clutch pedal is released. Who is right?
    A. A only
    B. B only
    C. Both A and B
    D. Neither A nor B (A.4)

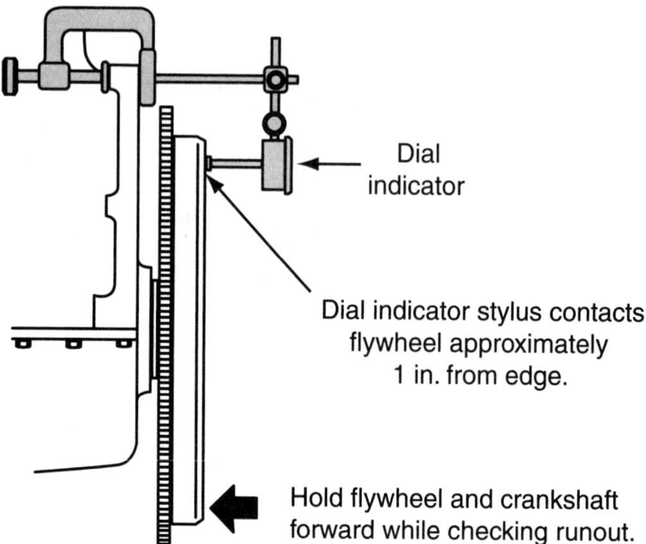

51. With the dial indicator positioned as shown in the figure, the measurement being performed is:
    A. crankshaft end play.
    B. crankshaft warpage.
    C. rear main bearing wear.
    D. rear engine block alignment. (A.9)
52. Technician A says you can test some vehicle speed sensors by checking with an ohmmeter. The meter should indicate a specified resistance to determine if it is good or bad. Technician B says when you check a vehicle speed sensor signal, you should use an oscilloscope as the drive gear is rotated. Who is right?
    A. A only
    B. B only
    C. Both A and B
    D. Neither A nor B (C.13)
53. A vacuum-shifted four-wheel drive (4WD) system does not shift into 4WD. Technician A says the engine vacuum may be low. Technician B says the vacuum motor at the front axle may be the problem. Who is right?
    A. A only
    B. B only
    C. Both A and B
    D. Neither A nor B (F.1)
54. Technician A says an improper shift linkage adjustment may cause hard transaxle shifting. Technician B says an improper shift linkage adjustment may cause the transaxle to stick in gear. Who is right?
    A. A only
    B. B only
    C. Both A and B
    D. Neither A nor B (C.2)

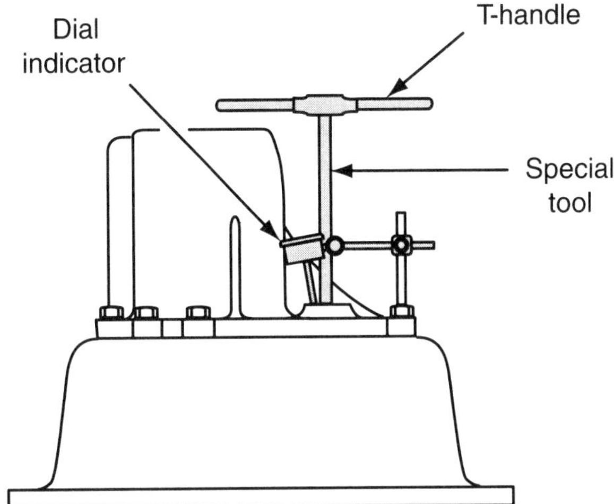

55. When taking the measurement shown in the figure:
    A. a new bearing cup is installed with a shim in the clutch bell housing side of the transaxle case.
    B. a medium load should be applied to the differential in the upward direction.
    C. the proper shim thickness is equal to the differential end play recorded on the dial indicator.
    D. the bolts between the transaxle case halves must be tightened to one-half the specified torque. (C.18)
56. All of the following statements regarding four-wheel drive (4WD) front-drive axles and joints are true EXCEPT:
    A. The inner tripod joint is held on the axle shaft with a snap ring.
    B. A special swaging tool may be required to tighten the outer boot clamps.
    C. The new joint is coated with the grease supplied, and the remaining grease is discarded.
    D. A worn CV joint may cause a clicking noise while cornering. (F.6)
57. A light-duty, rear-wheel drive truck has a growling noise that is worse during acceleration and deceleration. Technician A says an outer rear axle wheel bearing may be rough and worn. Technician B says the driveshaft center support bearing may be rough and worn. Who is right?
    A. A only
    B. B only
    C. Both A and B
    D. Neither A nor B (D.3)
58. Bearing preload is:
    A. to pack a bearing before installation.
    B. the amount of pressure applied to a bearing while the transmission is under load.
    C. the amount of pressure applied to a bearing upon assembly of the transmission.
    D. not adjustable. (B.13)

59. Technician A says a plugged transfer case vent may cause seal leakage. Technician B says that a remote transfer case vent helps to prevent moisture from entering the transfer case when driving through water. Who is right?
   A. A only
   B. B only
   C. Both A and B
   D. Neither A nor B (F.4)

60. Technician A says in four-wheel drive (4WD) low, the power flow in the transfer case goes from the input shaft through the sun gear and planetary carrier to provide a gear reduction. Technician B says in 4WD low, the annulus gear in the planetary gear is rotating counterclockwise. Who is right?
   A. A only
   B. B only
   C. Both A and B
   D. Neither A nor B (F.2)

61. While discussing a manual transmission that jumps out of second gear, Technician A says there may be excessive end play between the second-speed gear and its matching synchronizer. Technician B says the detent springs on the shift rail may be weak. Who is right?
   A. A only
   B. B only
   C. Both A and B
   D. Neither A nor B (B.1)

62. Technician A says that on some vehicles, the transmission and transfer case are removed as one assembly. Technician B says transfer cases may be removed as separate assemblies because the cases are made of cast-iron and the transmission and transfer case assembly would be too heavy for a transmission jack. Who is right?
   A. A only
   B. B only
   C. Both A and B
   D. Neither A nor B (F.3)

63. A limited slip differential chatters while cornering. Technician A says the differential may be filled with the wrong lubricant. Technician B says friction and steel plates may be worn and burned. Who is right?
   A. A only
   B. B only
   C. Both A and B
   D. Neither A nor B (E.1.8)

64. Technician A says if too much material is removed during flywheel resurfacing, the torsion springs on the clutch disc may contact the flywheel bolts, resulting in noise while engaging and disengaging. Technician B says if excessive material is removed when the flywheel is resurfaced, the slave cylinder rod may not have enough travel to release the clutch properly. Who is right?
   A. A only
   B. B only
   C. Both A and B
   D. Neither A nor B (A.7)

65. A worn pilot bearing may cause a rattling and growling noise while:
    A. the engine is idling and the clutch pedal is fully depressed.
    B. decelerating in high gear with the clutch pedal released.
    C. accelerating in low gear with the clutch pedal released.
    D. the engine is idling in neutral with the clutch pedal released. (A.6)

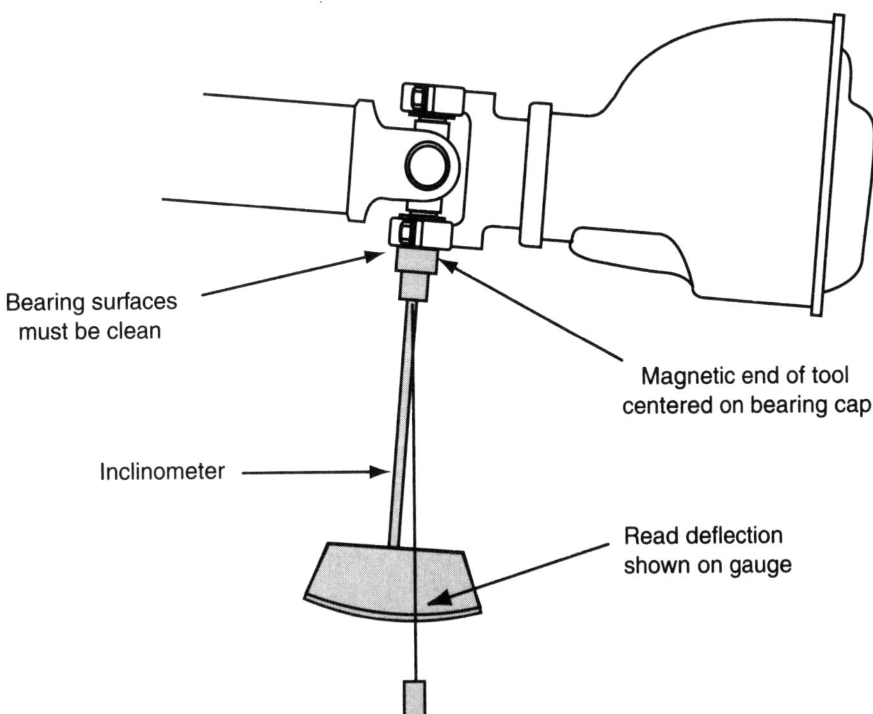

66. While discussing driveshaft and universal joint (U-joint) diagnosis in rear-wheel drive vehicles, Technician A says a worn U-joint may cause a squeaking noise that decreases in relation to vehicle speed. Technician B says a heavy vibration that only occurs during acceleration may be caused by a worn centering ball and socket on a double Cardan U-joint. Who is right?
    A. A only
    B. B only
    C. Both A and B
    D. Neither A nor B (D.1)

67. Technician A says an axle shaft with excessive runout may cause a vibration while driving at a constant speed of 60 mph (97 km/h). Technician B says an axle shaft with excessive runout may cause a vibration only while accelerating at low speed. Who is right?
    A. A only
    B. B only
    C. Both A and B
    D. Neither A nor B (E.4.1)

68. All of the following problems may cause a vibration on a four-wheel drive (4WD) vehicle that is more noticeable when changing throttle position EXCEPT:
    A. worn U-joints.
    B. worn front-drive axle joints.
    C. incorrect driveshaft angles.
    D. worn driveshaft slip joint splines. (F.6)

69. Technician A says the pinion bearings should be lubricated when the pinion turning torque is measured. Technician B says the pinion nut may be loosened to obtain the specified turning torque. Who is right?
    A. A only
    B. B only
    C. Both A and B
    D. Neither A nor B (E.1.5)
70. The most likely cause of clutch chatter among the following is:
    A. a worn, rough clutch release bearing.
    B. a worn, rough pilot bearing.
    C. excessive input shaft end play.
    D. weak clutch plate torsional springs. (A.1)
71. Technician A says an accurate differential case runout measurement may be performed with scored side bearings. Technician B says the ring gear runout should be measured before the case runout. Who is right?
    A. A only
    B. B only
    C. Both A and B
    D. Neither A nor B (E.1.7)

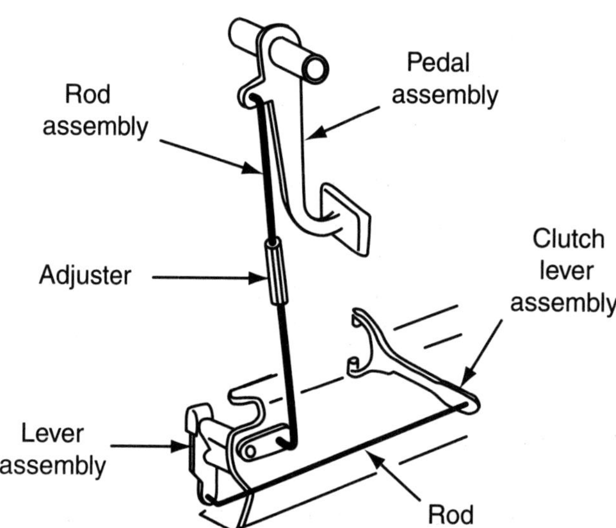

72. As shown in the figure, lack of clutch pedal free play may cause:
    A. hard shifting.
    B. incomplete clutch release.
    C. transaxle gear damage.
    D. clutch slipping. (A.2)

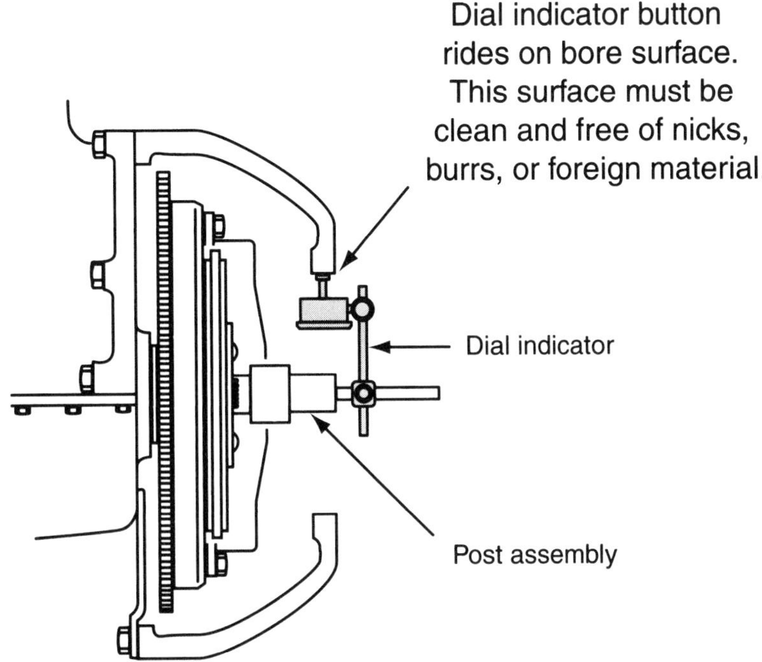

73. Excessive misalignment between the bell housing and the engine block would most likely cause:
    A. reduced clutch pedal free play.
    B. a growling noise when the clutch pedal is depressed.
    C. a vibration at higher speeds.
    D. clutch grabbing and chatter. (A.8)

74. A rear-wheel drive vehicle has a vibration that increases in relation to vehicle speed. Technician A says the balance pad may have fallen off the drives haft. Technician B says some of the wheels may be out of balance. Who is right?
    A. A only
    B. B only
    C. Both A and B
    D. Neither A nor B (D.6)

75. While discussing a four-speed manual transaxle that jumps out of third gear, Technician A says the shift rail detent spring tension on the 3-4 shift rail may be weak. Technician B says there may be excessive wear on the fourth-speed gear dog teeth. Who is right?
    A. A only
    B. B only
    C. Both A and B
    D. Neither A nor B (C.6)

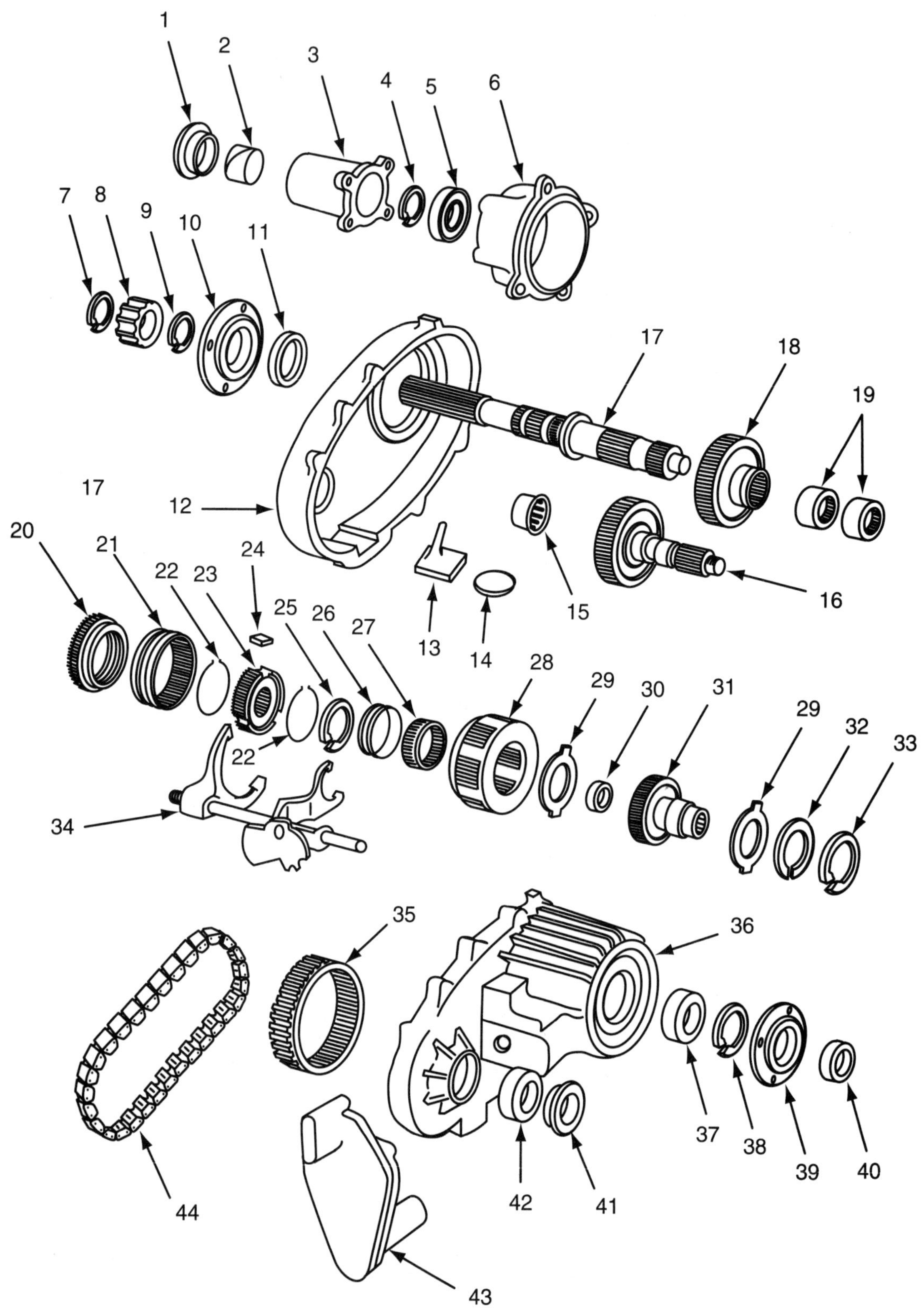
1
2
3
4
5
6
7
8
9
10
11
17
18
19
17
12
20
21
22
23
24
15
16
13
14
25
26
27
28
29
30
31
29
32
33
22
34
35
36
37
38
39
40
42
41
44
43

1. Rear output shaft seal	23. Sychronizer hub
2. Extension housing bushing	24. Sychronizer struts (3)
3. Extension housing	25. Snap ring
4. Retainer	26. Range shift hub
5. Rear output shaft bearing	27. Range gear
6. Pump retaininer hiusing	28. Planetary carrier
7. Tone wheel retainer	29. Thrust washer (2)
8. Tone wheel	30. Mainshaft pilot bearing
9. Tone wheel retainer	31. Input gear
10. Oil pump	32. Carrier lock ring
11. Oil pump seal	33. Snap ring
12. Rear case half	34. Shifting fork mechanism
13. Pump pick-up screen	35. Annulus gear
14. Magnet	36. Front case half
15. Front output rear bearing	37. Input bearing
16. Front output (driven) shaft	38. Snap ring
17. Main shaft	39. Input bearing retainer
18. Drive sprocket	40. Input bearing seal
19. Drive sprocket bearings	41. Front output shaft seal
20. Main drive sychronizer ring	42. Front output shaft bearing
21. Sychronizer sleeve	43. Encoder motor
22. Sychronizer strut spring (2)	44. Drive chain

76. All of the following are true about transfer case inspection and assembly EXCEPT:
    A. The transfer case chain should be inspected for stretching and looseness.
    B. Thrust washer thickness should be measured to check for wear or for select fit.
    C. Specification measurements should have been taken during disassembly.
    D. All parts should be cleaned before installation and assembled dry. (F.2)

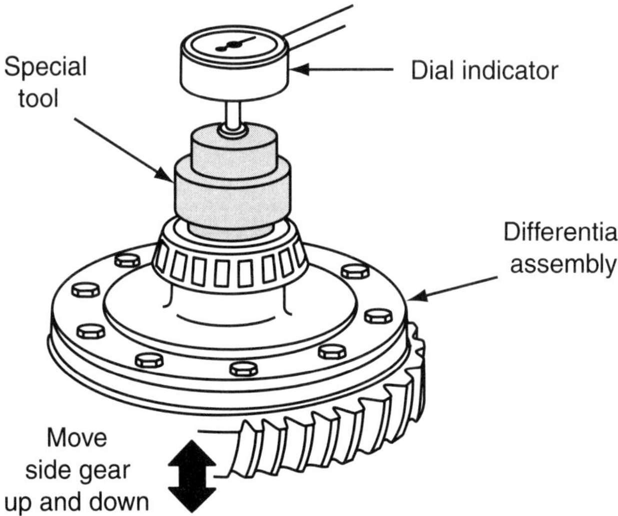

77. While determining the proper differential side gear thrust washer thickness:
    A. measure the end play only on one side gear to calculate the side gear spacer washer thickness.
    B. measure the side gear end play with the thrust washers behind the gears.
    C. the correct thickness of the side gear thrust washer provides the specified side gear end play.
    D. the correct thickness of the side gear thrust washer provides a slight side gear preload. (C.16)

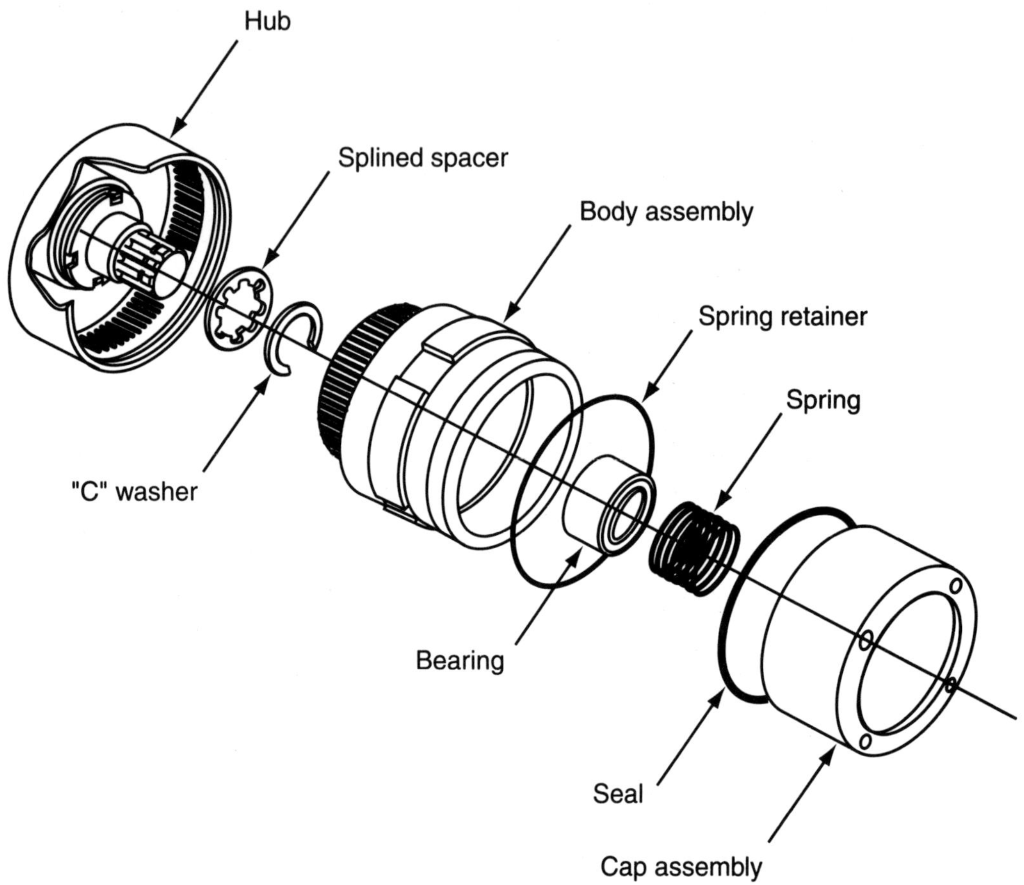

78. Technician A says automatic locking hubs should be packed with grease. Technician B says the cap on automatic locking hubs should be packed with grease. Who is right?
    A. A only
    B. B only
    C. Both A and B
    D. Neither A nor B (F.8)
79. Technician A says shims may be installed between the bell housing and engine block mating surfaces to correct bell housing face runout. Technician B says excessive bell housing face runout may be caused by overheated clutch disc facing. Who is right?
    A. A only
    B. B only
    C. Both A and B
    D. Neither A nor B (A.10)
80. Clutch chatter could most likely be caused by:
    A. excessive crankshaft end play.
    B. loose engine main bearings.
    C. a badly scored pressure plate.
    D. an improper pressure plate-to-flywheel position. (A.5)

81. A front axle on a four-wheel drive (4WD) vehicle is leaking oil. Technician A says that the axle vent may be plugged. Technician B says that the axle may be overfilled with lubricant.
    A. A only
    B. B only
    C. Both A and B
    D. Neither A nor B (F.9)
82. To measure differential case runout, which of the following tools would be used?
    A. Torque wrench
    B. Dial indicator
    C. Feeler gauge
    D. Micrometer (E.2.5)
83. When replacing a differential, Technician A says that the differential bearings can be used with the new differential. Technician B says that if the differential bearings are replaced, the bearing races must be replaced. Who is right?
    A. A only
    B. B only
    C. Both A and B
    D. Neither A nor B (E.2.2)

6 Additional Test Questions for Practice

Additional Test Questions

Please note the letter and number in parentheses following each question. They match the overview in section 4 that discusses the relevant subject matter. You may want to refer to the overview using this cross-referencing key to help with questions posing problems for you.

1. Technician A says that transmissions with internal linkage have no internal adjustment. Technician B says that only transmissions with external linkage can be adjusted. Who is right?
   A. A only
   B. B only
   C. Both A and B
   D. Neither A nor B (B.2)

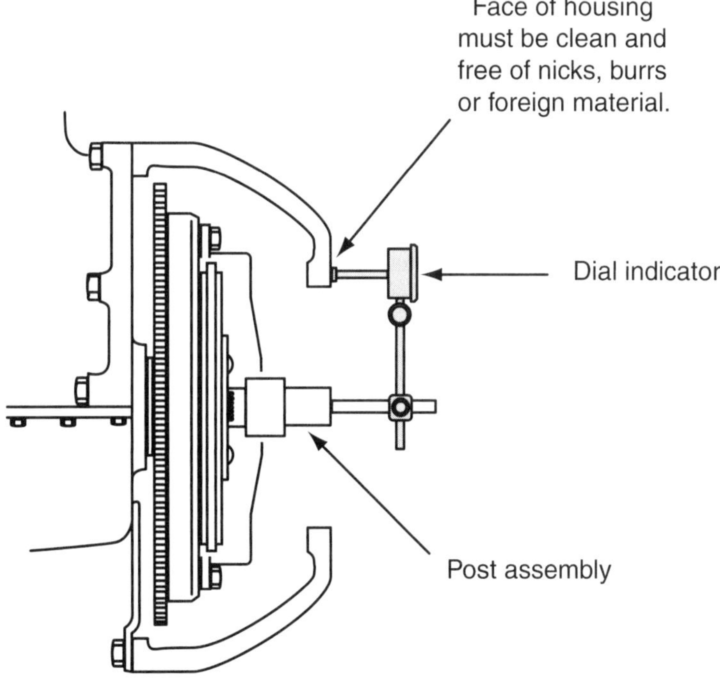

2. To correct excessive runout, as shown on the dial indicator:
   A. replace the engine mounts.
   B. adjust eccentric bell housing dowels.
   C. install bell housing shims.
   D. replace the clutch disc. (A.10)

3. To measure axle shaft end play:
   A. the differential cover must be removed.
   B. the vehicle needs to be in neutral.
   C. the brake drum needs to be removed.
   D. the axle shaft needs to be removed from the axle housing. (E.4.4)
4. Differential fluid is leaking from an axle housing. Which of the following would be the LEAST likely cause?
   A. The axle housing vent is plugged.
   B. The differential is overfilled.
   C. The axle shaft bearings are worn.
   D. The wrong fluid was used and is too thin. (E.1.1)
5. A bearing-type noise begins to come from the clutch and transmission area of a vehicle just as the clutch is almost completely disengaged. There is no noise when the clutch pedal is initially depressed. Technician A says that the clutch pilot bearing may be worn out. Technician B says that the transmission input shaft bearing may be faulty. Who is right?
   A. A only
   B. B only
   C. Both A and B
   D. Neither A nor B (A.1)
6. While inspecting a reverse idler gear from a manual transmission, Technician A says that you should check the center bore for a smooth mar-free surface. Technician B says that the reverse idler gear is splined in the center bore and that the splines should be checked for excessive wear or damage. Who is right?
   A. A only
   B. B only
   C. Both A and B
   D. Neither A nor B (B.11)
7. An extension housing has burrs and gouges on the mating surface. Technician A says that if they are not excessive, they can be repaired with a file. Technician B says that the mating surfaces are machined surfaces and they should be filed with fine grit sandpaper. Who is right?
   A. A only
   B. B only
   C. Both A and B
   D. Neither A nor B (B.14)
8. During disassembly, mark the position of the flywheel in relation to the:
   A. transmission input shaft.
   B. crankshaft.
   C. clutch disc.
   D. bell housing. (A.9)
9. To measure driveshaft runout, you should do all of the following EXCEPT:
   A. place the vehicle transmission in neutral.
   B. use a magnetic-base dial indicator.
   C. roll the driveshaft on a flat surface to check for damage.
   D. clean the driveshaft surface for an accurate runout check. (D.5)

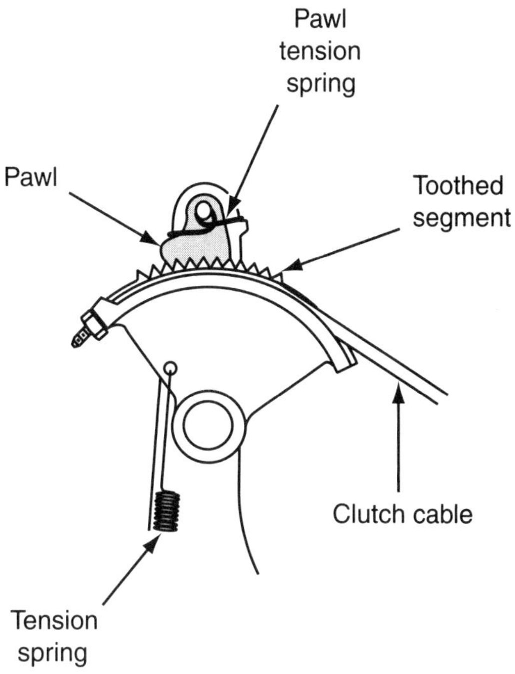

10. A clutch with a self-adjusting cable has:
    A. 1 inch (25.4 mm) of clutch pedal free play.
    B. 2 inches (50.8 mm) of clutch pedal free play.
    C. a constant running release bearing.
    D. an overcenter assist spring. (A.2)
11. When installing a replacement U-joint that has a grease fitting, the fitting should point toward the:
    A. front of the vehicle.
    B. transmission.
    C. differential.
    D. driveshaft. (E.1.2)
12. When removing any transaxle from a vehicle, you will need to:
    A. install an engine support.
    B. drain the engine oil.
    C. disconnect the positive battery cable.
    D. remove the engine. (C.4)
13. Excessive noise coming from the transfer case may be caused by all of the following EXCEPT:
    A. low fluid level.
    B. misalignment of the transfer chain.
    C. a worn universal joint.
    D. a damaged output shaft bearing. (F.1)
14. A remote vent on a differential is used to:
    A. increase pressure in the differential.
    B. keep moisture out of the differential.
    C. keep lubricant from coming out of the differential.
    D. add lubricant to the differential. (F.9)

15. A manual-shift transfer case will not shift into four-wheel drive (4WD). The cause of this problem could be:
    A. the fluid is low in the transfer case.
    B. the front driveshaft universal joints are bound up.
    C. the shift linkage needs lubrication.
    D. the electronic shift motor is bad. (F.5)
16. When removing the transaxle differential from the vehicle, the LEAST likely component to be removed would be the:
    A. axle shaft.
    B. transaxle.
    C. engine.
    D. lower control arms. (C.15)
17. If an axle shaft has seal surface damage, all of the following should be inspected EXCEPT:
    A. axle seal.
    B. axle bearing.
    C. rear axle housing.
    D. brake drum. (E.4.3)

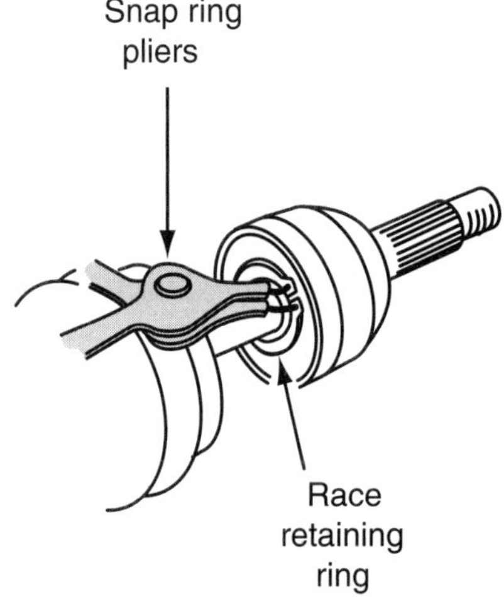

18. Technician A says that a clunking noise while turning could be caused by a bad constant-velocity (CV) joint. Technician B says a CV joint could also make noise when the vehicle is traveling straight. Who is right?
    A. A only
    B. B only
    C. Both A and B
    D. Neither A nor B (D.2)
19. Technician A says that if the differential is overfilled, it will cause axle seals to leak. Technician B says that worn axle bearings will cause leaks. Who is right?
    A. A only
    B. B only
    C. Both A and B
    D. Neither A nor B (E.1.1)

20. A broken detent spring could cause any of the following problems EXCEPT:
    A. the transmission to jump out of gear.
    B. harsh shifting.
    C. lock up between two gears.
    D. a growling noise. (C.6)
21. While discussing reverse idler gears in a manual transmission, Technician A says that a reverse idler gear may ride on needle bearings. Technician B says that a reverse idler gear may ride on a bronze bushing. Who is right?
    A. A only
    B. B only
    C. Both A and B
    D. Neither A nor B (C.10)
22. The pinion gear rotating torque is too high. To fix this problem:
    A. the bearings will have to be replaced.
    B. the pinion gear will have to be replaced.
    C. the pinion nut can be slightly loosened.
    D. the collapsible spacer must be replaced and the turning torque reset. (E.2.3)

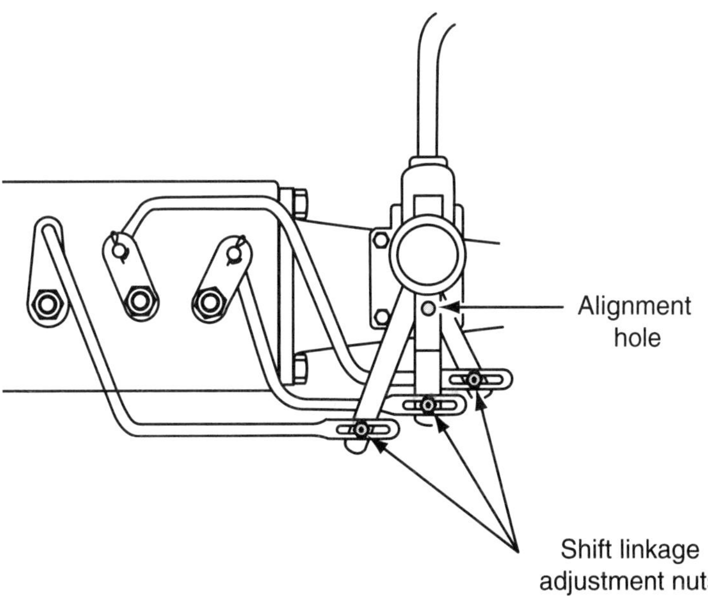

23. When replacing transmission shift levers as shown in the figure, Technician A says that all lever bushings should be replaced. Technician B says that only the bushings that are worn should be replaced. Who is right?
    A. A only
    B. B only
    C. Both A and B
    D. Neither A nor B (B.2)
24. During a transmission inspection, a damaged input bearing is found. Which of the following is also likely to be damaged?
    A. The bearing retainer and seal
    B. The counter shaft bearing
    C. 2nd gear
    D. The counter gear (A.8)

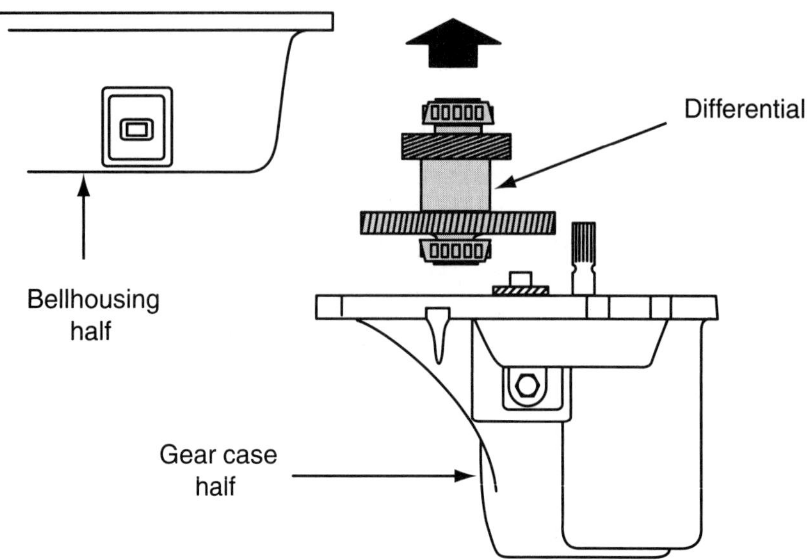

25. In order to remove the differential assembly in a transaxle, you must:
    A. remove the transaxle assembly.
    B. split the transaxle case while it is still in the vehicle.
    C. remove the engine from the vehicle.
    D. disassemble the entire transaxle. (C.15)
26. Technician A says that the speedometer drive gear does not have to be replaced when a new speedometer cable core is replaced. Technician B says that by changing the drive gear, the speedometer readings may change. Who is right?
    A. A only
    B. B only
    C. Both A and B
    D. Neither A nor B (C.12)
27. When replacing an extension housing seal, which of the following does not need to be inspected?
    A. The extension housing bushing
    B. The front driveshaft yoke
    C. The extension housing gasket
    D. The input bearing (B.3)
28. While discussing the installation of a new pilot bushing or bearing, Technician A says that when you install a pilot bushing, you should lubricate the bushing with motor oil. Technician B says when you install a pilot bearing, you should lubricate the bearing with wheel bearing grease. Who is right?
    A. A only
    B. B only
    C. Both A and B
    D. Neither A nor B (A.6)

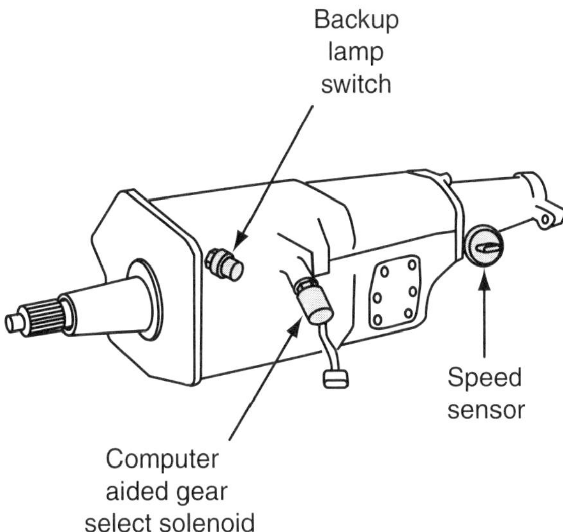

29. Technician A says that when removing a manual transmission, always check for a rear main engine seal leak. Technician B says that the engine does not have to be supported when the transmission is removed from the vehicle. Who is right?
    A. A only
    B. B only
    C. Both A and B
    D. Neither A nor B (B.4)
30. All of the following can cause axle shaft seal leakage EXCEPT:
    A. worn axle shaft bearings.
    B. plugged axle vent.
    C. scored axle shafts.
    D. heat from the rear brakes. (E.4.3)
31. Technician A says that the blocking ring with the sharp edges is good. Technician B says that the sharp edges will prevent the transmission from jumping out of gear. Who is right?
    A. A only
    B. B only
    C. Both A and B
    D. Neither A nor B (B.9)
32. Excessive end play in the counter gear could cause damage to all of the following EXCEPT the:
    A. transmission case.
    B. gear teeth.
    C. counter shaft.
    D. release bearing. (B.10)
33. A new universal joint has been installed in a vehicle. Technician A says the universal joint should be greased until grease comes out of the four caps. Technician B says that the universal joint is prelubricated, and that driveshaft balance affects the amount of grease to install. Who is right?
    A. A only
    B. B only
    C. Both A and B
    D. Neither A nor B (D.2)

34. All of the following statements about ring gears are true EXCEPT:
    A. Hunting-type ring and pinion gear sets must be timed.
    B. Loose ring gear bolts may cause a gear chuckle or knocking noise while driving the vehicle.
    C. Damaged ring gear and pinion gear teeth may cause a ticking noise while driving the vehicle.
    D. The grooved, painted tooth on the pinion gear must be meshed with the painted, notched ring gear teeth on some ring gearsets. (E.1.3)
35. The backup lights are staying on in all gears. What is the LEAST likely cause?
    A. A short in the backup light switch
    B. The linkage out of adjustment
    C. A short in the wiring harness
    D. A bad brake light switch (B.16)
36. Technician A says constant-running release bearings are used with hydraulically controlled clutches. Technician B says release bearings move away from the pressure plate to disengage the clutch. Who is right?
    A. A only
    B. B only
    C. Both A and B
    D. Neither A nor B (A.4)
37. When checking flywheel runout, what should be checked first?
    A. The ring gear
    B. The crankshaft end play
    C. The pilot bushing
    D. Free travel in the clutch pedal (A.7)
38. A constant whining noise is coming from the differential of a vehicle. Which of the following could be the cause?
    A. The preload and backlash are not set properly.
    B. The wrong differential lube was used.
    C. The side gears are damaged.
    D. The spider gears are damaged. (C.14)
39. On a vehicle with a manual transaxle that jumps out of second gear, all of the following could be the cause EXCEPT:
    A. a worn second gear blocking ring.
    B. an excessive main shaft end play.
    C. a shifter linkage out of adjustment.
    D. a worn ring gear. (C.2)

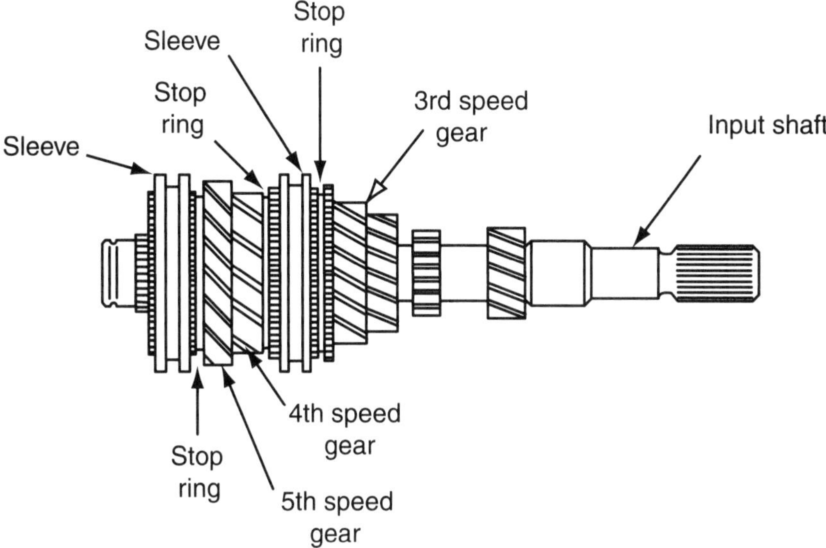

40. The clearance on the fourth-speed gear is less than specified. Technician A says this may result in noise while driving in fourth gear. Technician B says this problem may cause hard shifting into fourth gear. Who is right?
    A. A only
    B. B only
    C. Both A and B
    D. Neither A nor B (B.9)
41. The speedometer drive gear is all of the following EXCEPT:
    A. mounted on the output shaft.
    B. made of a plastic nylon material.
    C. machined into the output shaft.
    D. a gear with helical teeth. (C.12)
42. While discussing the main shaft third gear clutch teeth, Technician A says the teeth should be rounded. Technician B says the teeth should have a sharp, beveled edge. Who is right?
    A. A only
    B. B only
    C. Both A and B
    D. Neither A nor B (B.9)
43. When removing a transfer case from the vehicle, which of the following components listed is LEAST likely to be disconnected or removed?
    A. The front prop shaft
    B. The transmission
    C. The rear prop shaft
    D. The linkage (F.3)
44. Technician A says that a worn pilot bearing or bushing is a common result of a misalignment condition. Technician B says that a bent transmission input shaft or a bent clutch disc may cause a misalignment condition. Who is right?
    A. A only
    B. B only
    C. Both A and B
    D. Neither A nor B (A.8)

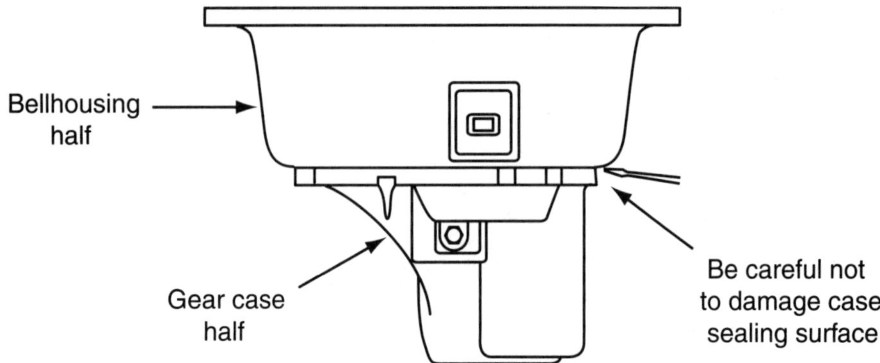

45. While discussing manual transaxles, Technician A says that most transaxle cases are sealed with RTV sealant. Technician B says that some transaxle cases have paper gaskets. Who is right?
    A. A only
    B. B only
    C. Both A and B
    D. Neither A nor B (C.3)

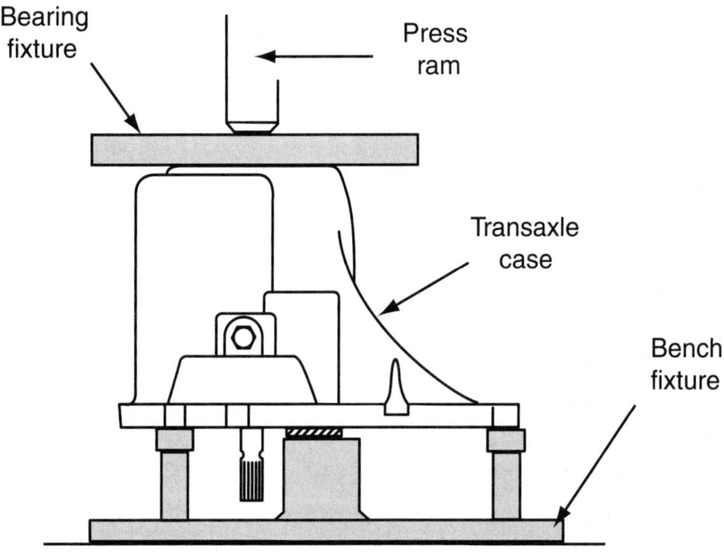

46. After removal of a transaxle, all of the following components will require opening the transaxle case before removal EXCEPT the:
    A. input shaft.
    B. output shaft.
    C. pilot bearing.
    D. differential bearings. (C.5)

47. Technician A says 10W-30 oil can be used in a limited slip differential. Technician B says that using the wrong type of oil may cause chatter when the vehicle is cornering. Who is right?
    A. A only
    B. B only
    C. Both A and B
    D. Neither A nor B (E.1.8)
48. Which of the following is the best way to check component wear when assembling a transfer case?
    A. Prelube the components
    B. Clean components thoroughly
    C. Clean all sealing surfaces
    D. Measure clearances (F.2)
49. When adjusting the pinion depth on the ring gear, Technician A says to replace the collapsible spacer. Technician B says that pinion depth can also be adjusted by installing a selective pinion bearing race. Who is right?
    A. A only
    B. B only
    C. Both A and B
    D. Neither A nor B (E.1.4)
50. Technician A says that when installing a cork gasket, no added sealant is required. Technician B says that rubber gaskets should be installed without any added sealant. Who is right?
    A. A only
    B. B only
    C. Both A and B
    D. Neither A nor B (B.3)
51. When draining the fluid from a manual transaxle, a gold-colored material is seen in the fluid. The LEAST likely cause is a worn:
    A. blocking ring.
    B. second gear.
    C. thrust washer.
    D. shift fork. (C.19)
52. Technician A says if side gears or spider gears are damaged, noise would only be heard when the vehicle is turned. Technician B says that side gears are shimmed for proper clearance. Who is right?
    A. A only
    B. B only
    C. Both A and B
    D. Neither A nor B (C.14)
53. All of the following are true about removing a differential assembly EXCEPT:
    A. the axle shafts must be removed.
    B. the bearing caps should be marked to the housing.
    C. the bearing races and shim pack should not be mixed up.
    D. the pinion gear always stays in the axle housing. (E.2.2)

54. When measuring pinion bearing preload:
    A. the axle must be fully assembled.
    B. the axle shafts must be removed.
    C. the differential case must be removed.
    D. a spring scale with a hook can be used to measure the preload. (E.1.2)

55. Technician A says that a transmission that is hard to shift may have a problem with the linkage not being lubricated. Technician B says that this problem may be caused by too strong of a pressure plate installed in the vehicle's clutch system. Who is right?
    A. A only
    B. B only
    C. Both A and B
    D. Neither A nor B (B.1)

56. A clutch makes a loud "chirping" noise when the clutch pedal is depressed with the engine running. Technician A says that the release bearing is worn and must be replaced. Technician B says that the input shaft bearing is worn and must be replaced. Who is right?
    A. A only
    B. B only
    C. Both A and B
    D. Neither A nor B (A.1)

57. The speedometer driving gear is located on the:
    A. vehicle speed sensor.
    B. main shaft.
    C. output shaft.
    D. axle shaft. (B.15)

58. Technician A says that when checking rear axle runout, rotate the axle slowly. Technician B says to use a dial indicator to measure runout. Who is right?
    A. A only
    B. B only
    C. Both A and B
    D. Neither A nor B (E.4.4)

59. The LEAST likely cause of clutch chatter would be:
    A. worn input shaft splines.
    B. worn torsional springs.
    C. a worn pilot bearing.
    D. a bent clutch disc. (B.7)

60. To measure the axle shaft end play on the front axles of a 4WD vehicle, all of the following must be removed EXCEPT the:
    A. lug nuts.
    B. brake drums.
    C. brake pads or shoes.
    D. wheel and tire. (E.4.4)

61. A transfer case is considered full when:
    A. fluid comes out of the front input when the yoke is removed.
    B. fluid is at the halfway point of the sight glass window.
    C. fluid just starts to come back out of the fill hole when the plug is removed.
    D. at least 4 inches (102 mm) of fluid level is found when a piece of mechanic's wire is inserted into the fill hole. (E.3.2)

62. After inspecting the ring gear, a toe pattern is found. Which of the following is the most likely cause?
    A. A worn axle bearing
    B. A worn collapsible pinion spacer
    C. Worn spider gears
    D. A worn differential bearing (E.1.6)

63. On a rear-wheel drive vehicle, the driveshaft is most likely balanced:
    A. before it is installed in the vehicle.
    B. after it is installed in the vehicle.
    C. with the vehicle moving.
    D. with the vehicle stopped. (D.4)

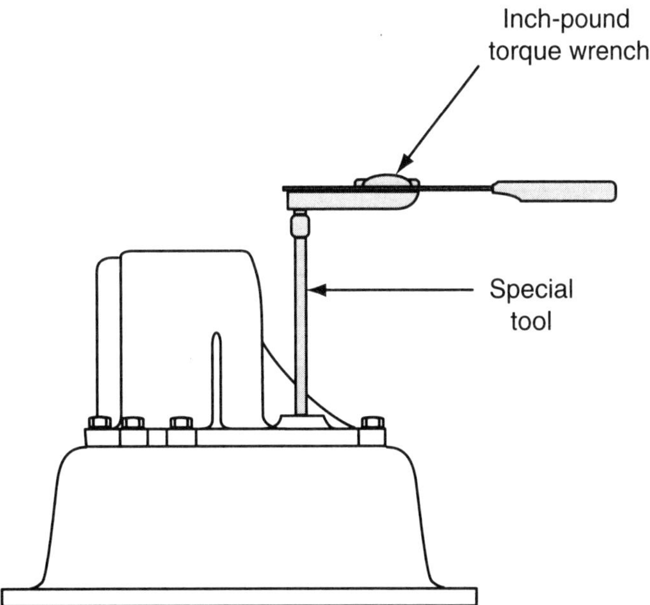

64. When the turning torque in the illustration shown is less than specified:
    A. a thicker shim should be installed behind the side bearing cup in the bell housing side of the case.
    B. a thinner shim should be installed behind both differential side bearing cups.
    C. a thinner shim should be installed behind both differential side bearings.
    D. a thicker shim should be installed behind both differential side gears. (C.18)

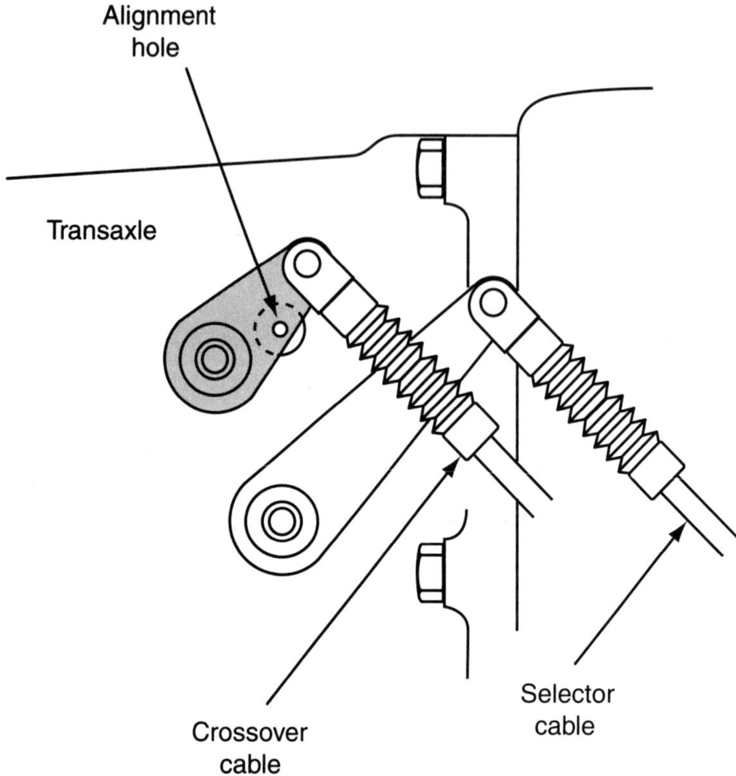

65. In the figure shown, the transaxle is in the neutral position to allow the technician to:
    A. replace the shift cables.
    B. adjust the shift cables.
    C. modify the shift cables.
    D. repair the transaxle case. (C.2)
66. The distance from the rivet heads to the clutch facing surface should be no less than:
    A. 0.005 inch (0.127 mm).
    B. 0.008 inch (0.203 mm).
    C. 0.012 inch (0.305 mm).
    D. 0.025 inch (0.638 mm). (A.5)

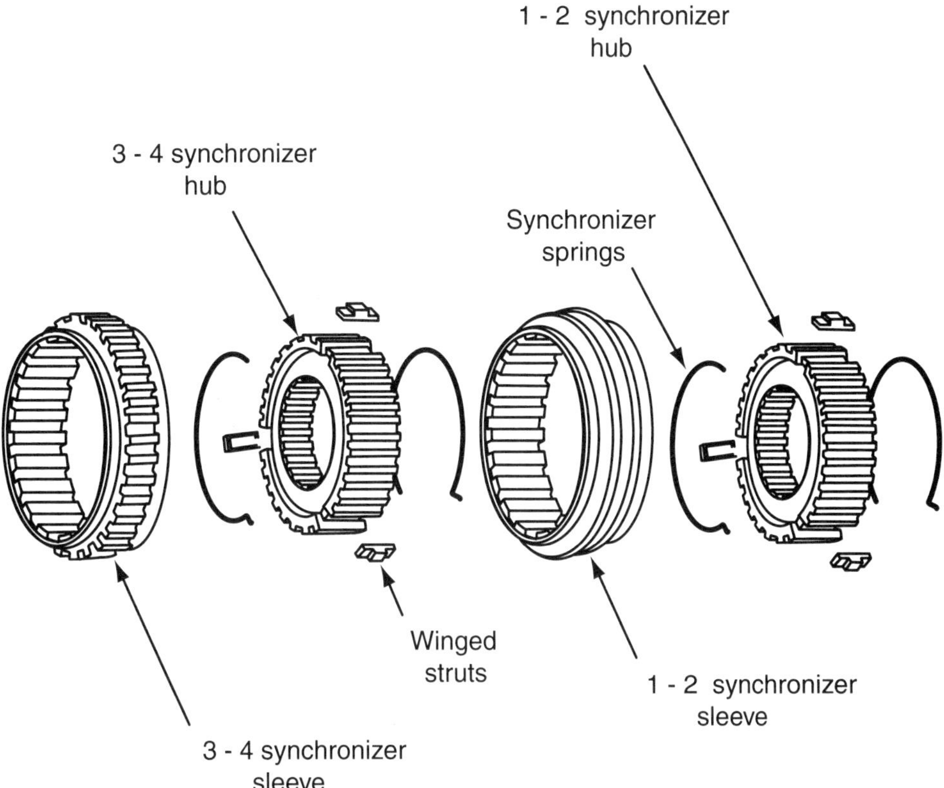

67. A shift fork is connected to the:
    A. forward and reverse gears.
    B. counter shaft.
    C. blocker rings.
    D. synchronizer assembly. (C.5)
68. A transmission mount saturated with oil should be:
    A. cleaned and reused.
    B. inspected for cracks.
    C. used for the rubber in the mount.
    D. discarded and replaced. (B.4)
69. The speedometer in a vehicle stopped working, but the sensor circuitry and the speedometer head are OK. Which of the following problems would be the LEAST likely cause?
    A. The drive and driven gears are stripped.
    B. The drive gear slips on the end of the sensor.
    C. The drive gear is not positioned correctly on the output shaft.
    D. The driven gear teeth are stripped. (C.13)
70. While discussing the major components of a hydraulic clutch linkage, Technician A says the master cylinder for the car's brakes is also used for the clutch. Technician B says the slave cylinder is connected to the clutch pedal and increases hydraulic pressure as the pedal is depressed. Who is right?
    A. A only
    B. B only
    C. Both A and B
    D. Neither A nor B (A.3)

71. While discussing the rebuilding of a transaxle, Technician A says that snap rings and spacers are usually replaced. Technician B says that snap rings and spacers usually cannot be obtained in a small kit, and must be obtained in a complete overhaul kit. Who is right?
    A. A only
    B. B only
    C. Both A and B
    D. Neither A nor B (C.8)
72. Technician A says that spider gears are used when the vehicle is cornering. Technician B says the individual spider gears rotate at different speeds when the vehicle is going straight. Who is right?
    A. A only
    B. B only
    C. Both A and B
    D. Neither A nor B (E.1.1)

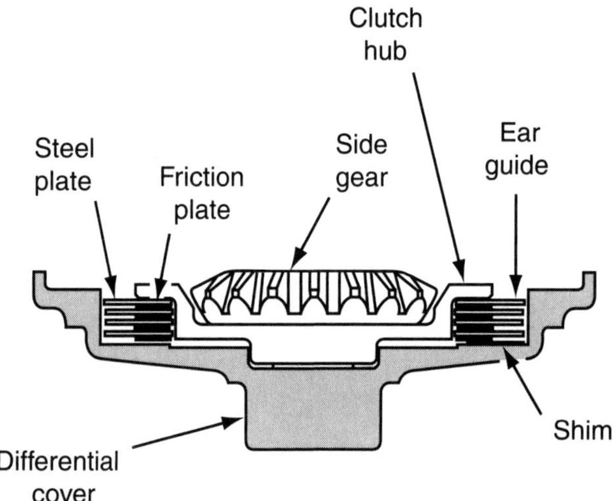

73. Technician A says the left-side differential bearing adjuster nut must be loosened to decrease ring gear backlash. Technician B says pinion preload can be adjusted with selective shims. Who is right?
    A. A only
    B. B only
    C. Both A and B
    D. Neither A nor B (E.1.6)
74. Most late-model manual transmissions are filled with:
    A. SAE 90 gear oil.
    B. power steering fluid.
    C. automatic transmission fluid or motor oil.
    D. a synthetic lubricant with a Teflon™ additive. (B.17)
75. A center shaft support bearing:
    A. should be greased like a universal joint.
    B. is usually a sealed bearing and is maintenance-free.
    C. is used on all rear-wheel drive vehicles.
    D. is part of the driveshaft and cannot be replaced separately. (D.3)

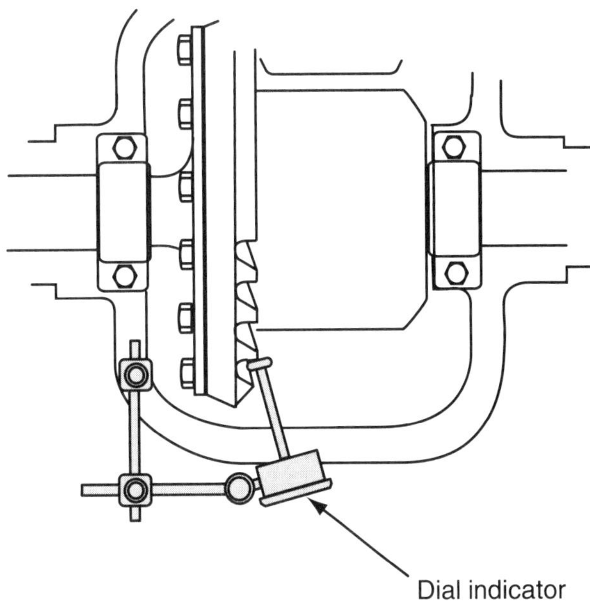

76. In the figure shown, a technician is measuring:
    A. ring gear runout.
    B. ring gear backlash.
    C. pinion gear backlash.
    D. bearing preload. (E.2.3)
77. When draining a limited slip differential, friction material is found in the fluid. What is the most likely cause?
    A. Damaged axle seals
    B. A damaged pinion seal
    C. Disc wear
    D. Damaged gaskets (C.19)
78. While discussing transaxle case mating surfaces, Technician A says that the mating surfaces do not have to be machined to precise specifications because the gasket or sealer will stop any leaks. Technician B says that a transaxle that has two halves to its case has a machined mating surface, but still requires a gasket or a sealant to prevent leakage. Who is right?
    A. A only
    B. B only
    C. Both A and B
    D. Neither A nor B (C.3)
79. When replacing an axle boot, which component is LEAST likely to be visually inspected?
    A. A CV joint
    B. A wheel bearing
    C. An axle shaft
    D. Axle seals (D.1)
80. Which tool is used to check ring gear runout?
    A. A feeler gauge
    B. A torque wrench
    C. A straightedge
    D. A dial indicator (E.1.3)

81. Too much driveline angle would most likely cause:
    A. binding universal joints.
    B. a loud humming noise.
    C. pinion gear damage.
    D. damage to the transmission mount. (D.6)
82. A vehicle with a manual transaxle is hard to shift on a cold start only. The cause of this problem could be:
    A. a rusted shift linkage.
    B. the clutch disc sticking to the flywheel.
    C. wrong lubricant in the transaxle.
    D. a bad slave cylinder. (C.19)
83. A growling noise is heard during deceleration. The LEAST likely cause of this could be:
    A. a worn pinion bearing.
    B. a worn side gear bearing.
    C. a worn axle bearing.
    D. worn spider gears. (E.4.1)
84. Which of the following is usually used to hold differential side bearings in place?
    A. Snap rings
    B. Lockwashers
    C. An interference fit
    D. Axle shaft splines (C.15)

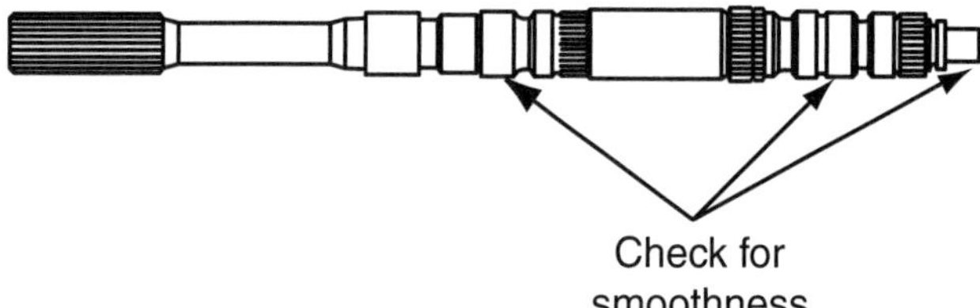

85. The figure shows a main shaft; the areas pointed out are:
    A. bearing surfaces.
    B. gear journals.
    C. oil journal locations.
    D. synchronizer mounting locations. (B.8)
86. Technician A says that preload on the pinion shaft should be measured with a dial indicator. Technician B says that preload can also be measured with a flat feeler gauge. Who is right?
    A. A only
    B. B only
    C. Both A and B
    D. Neither A nor B (C.17)

87. While discussing differential spider gears, Technician A says the spider gears that ride on the pinion shaft and the bore that runs through the gear should be smooth and shiny. Technician B says that the spider gears ride on the pinion shaft, but the bores in the gears have needle bearings. Who is right?
    A. A only
    B. B only
    C. Both A and B
    D. Neither A nor B (C.16)

88. Technician A says the driveshaft center support bearing can cause noises in neutral if the vehicle is stopped. Technician B says that some center support bearings need to be lubricated. Who is right?
    A. A only
    B. B only
    C. Both A and B
    D. Neither A nor B (D.3)

89. Technician A says that stripped threads in a transaxle case can be repaired with a helicoil. Technician B says that not all damaged threads in a transaxle should be repaired. Who is right?
    A. A only
    B. B only
    C. Both A and B
    D. Neither A nor B (C.11)

90. When measuring clearance between the blocking ring and the gear, you should use which of the following tools?
    A. A micrometer
    B. A dial indictor
    C. A feeler gauge
    D. A ruler (B.9)

91. After inspection of the reverse idler gear, Technician A says that since one tooth on the cluster gear is chipped, the entire gear must be replaced. Technician B says a reverse idler gear with a chipped tooth does not have to be replaced, because it is only used in reverse. Who is right?
    A. A only
    B. B only
    C. Both A and B
    D. Neither A nor B (B.11)

92. Technician A says that if the shaft is worn, the bushing inside the gear should be checked. Technician B says that if the shaft is worn, the case and all the reverse gears should be checked. Who is right?
    A. A only
    B. B only
    C. Both A and B
    D. Neither A nor B (C.10)

93. To check pinion flange runout:
    A. the driveshaft does not need to be removed.
    B. a straightedge and feeler gauge would be used.
    C. a dial indicator would be used.
    D. the pinion flange will need to be removed. (E.1.2)

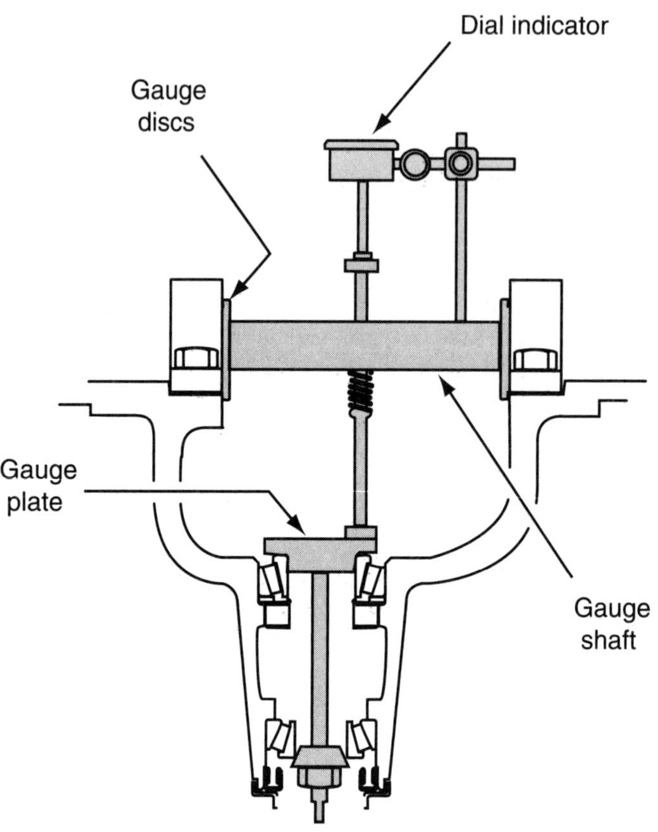

94. In the figure shown, after the dial indicator is rotated to the zero position, with the stem on the gauge plate, and then moved off the gauge plate, the dial indicator pointer moves 0.057 inches (1.45 mm) counterclockwise and the pinion gear is marked –4. The proper pinion depth shim is:
    A. 0.039 inch (0.99 mm).
    B. 0.041 inch (1.04 mm).
    C. 0.042 inch (1.07 mm).
    D. 0.043 inch (1.09 mm). (E.1.5)
95. To check the friction plates of a clutch pack:
    A. The friction plates should be measured with a micrometer.
    B. The friction plates should be measured with a feeler gauge.
    C. Only a visual inspection is necessary.
    D. The friction plates do not need to be removed from the clutch pack. (E.3.3)
96. When installing a new clutch disc, Technician A says that the torsional dampening springs on the clutch should face the flywheel. Technician B says that torsional dampening springs help smooth out engine pulsations. Who is right?
    A. A only
    B. B only
    C. Both A and B
    D. Neither A nor B (A.5)

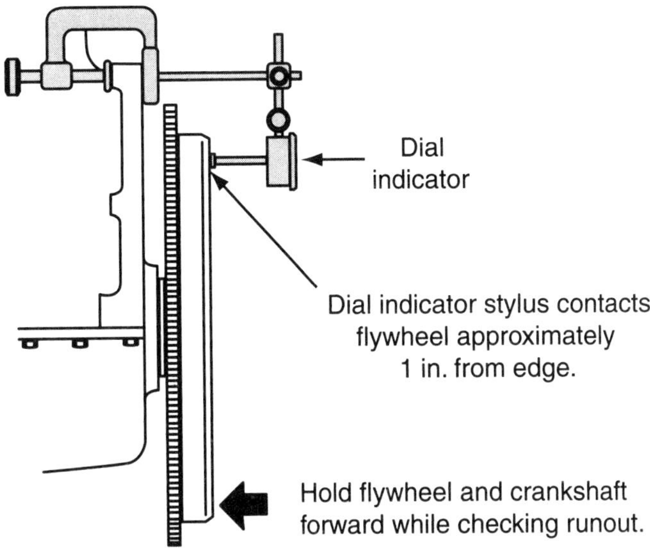

97. In the figure, Technician A says crankshaft end play can be checked with the tool in the above location. Technician B says that with the tool in this position you can check flywheel runout. Who is right?
    A. A only
    B. B only
    C. Both A and B
    D. Neither A nor B (A.9)

98. After inspecting a transmission mount and finding it oil soaked, Technician A says it should be replaced. Technician B says some mounts can be cleaned and reinstalled when oil soaked. Who is right?
    A. A only
    B. B only
    C. Both A and B
    D. Neither A nor B (A.11)

99. Which of the following is LEAST likely to be replaced when replacing the extension housing?
    A. The output shaft seal
    B. The tail housing gasket
    C. The speedometer O-ring
    D. The counter gear shaft (B.14)

100. Technician A says that the input shaft end play does not need to be checked before disassembling a transaxle. Technician B says to always rotate the input shaft to check turning effort before disassembling the transaxle. Who is right?
    A. A only
    B. B only
    C. Both A and B
    D. Neither A nor B (C.5)

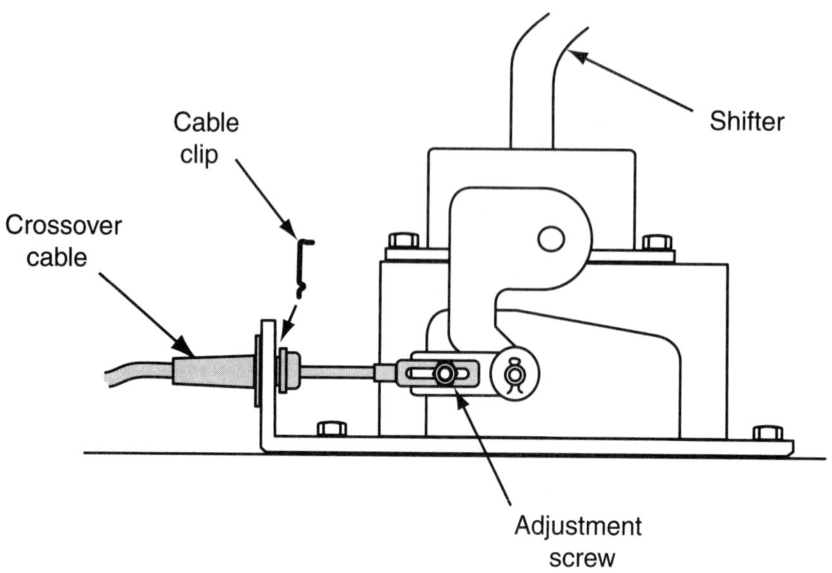

101. In the figure, Technician A says to check the selector cable if the vehicle has a sloppy shifter. Technician B says to check all external linkage bushings and grommets. Who is right?
   A. A only
   B. B only
   C. Both A and B
   D. Neither A nor B (C.2)

102. While driving a vehicle in four-wheel drive and under acceleration, a loud clicking noise is heard. Technician A says the front pinion gear is slipping on the ring gear in the axle housing. Technician B says that the noise could be the chain in the transfer case slipping because it is stretched. Who is right?
   A. A only
   B. B only
   C. Both A and B
   D. Neither A nor B (F.5)

103. Technician A says that a worn ring gear on the flywheel can cause clutch wear. Technician B says that a worn ring gear on the flywheel can affect the starter. Who is right?
   A. A only
   B. B only
   C. Both A and B
   D. Neither A nor B (A.7)

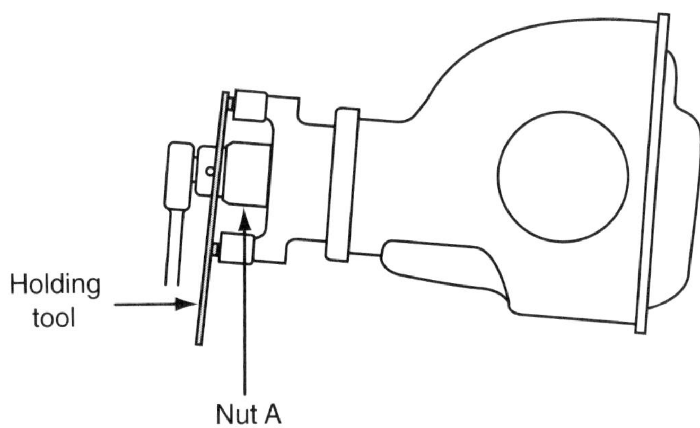

104. In the figure shown, if nut A is loose, which of the following is LEAST likely to happen?
   A. A clunking noise when vehicle is put in gear
   B. Excess pinion bearing wear
   C. A bad tooth pattern on the ring and pinion gears
   D. Wear on the spider gears (E.1.2)

105. When reassembling a transmission shaft, gear and synchronizer end play are set by using:
   A. a new main shaft.
   B. all new bearings.
   C. select-fit shims and thrust washers.
   D. all new snap rings. (B.12)

106. Technician A says that driveshaft runout should be checked with a digital micrometer set in the middle of the driveshaft. Technician B says that a dial indicator should be set at the differential end of the driveshaft to check runout. Who is right?
   A. A only
   B. B only
   C. Both A and B
   D. Neither A nor B (D.5)

107. Technician A says that when a wheel bearing is replaced, the bearing race should also be replaced. Technican B says the automatic locking hubs and the caps should be packed with grease. Who is right?
   A. A only
   B. B only
   C. Both A and B
   D. Neither A nor B (F.8)

108. When draining the manual transaxle oil, all of the following should be checked EXCEPT:
   A. metallic material in the fluid.
   B. the type of the oil and condition.
   C. leaks.
   D. filter condition. (C.17)

109. Technician A says that a worn blocker ring will cause the transaxle to have gear clash. Technician B says that a worn blocker ring will have sharp ridges in the cone area from slipping on the gear cone area. Who is right?
   A. A only
   B. B only
   C. Both A and B
   D. Neither A nor B (C.9)

110. An input shaft is supported by and uses:
   A. a tapered bearing.
   B. needle bearings.
   C. a sealed bearing.
   D. ball bearings. (C.7)

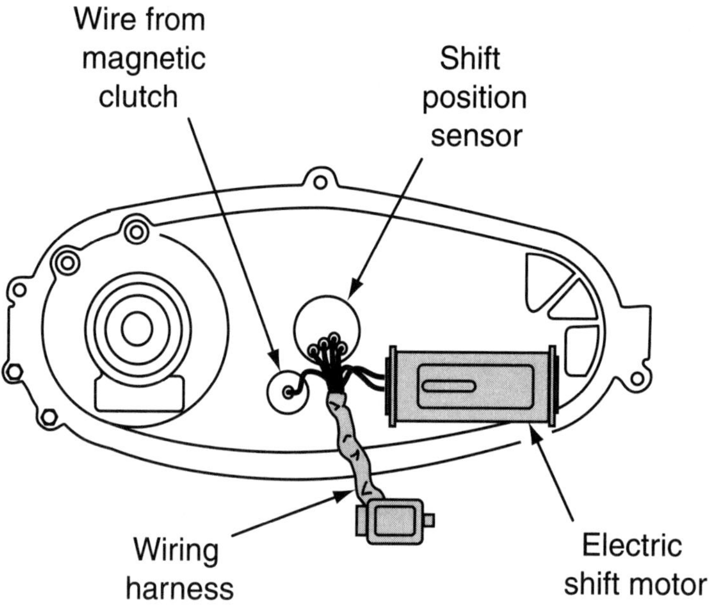

111. The best way to check mating surfaces for warpage on a transaxle case is to use a:
   A. straightedge.
   B. dial indicator.
   C. micrometer.
   D. flat surface. (C.3)

112. An electronic transfer case does not engage. Which of the following would be the LEAST likely cause?
   A. A bad electronic shift motor
   B. A blown fuse
   C. The 4WD engage switch
   D. A rusted linkage (F.10)

113. Technician A says that when filling a transfer case, fluid level can be checked by looking in the sight glass. Technician B says to fill the transfer case until it begins to come out of the fill vent. Who is right?
   A. A only
   B. B only
   C. Both A and B
   D. Neither A nor B (E.3.2)

114. When an axle shaft is removed from the differential it should be:
    A. stored standing straight up.
    B. laid flat.
    C. rolled on the ground to check for runout.
    D. left under the vehicle. (E.4.3)

115. Technician A says that a clutch may slip when it is out of adjustment. Technician B says that a transmission may grind when the clutch is out of adjustment. Who is right?
    A. A only
    B. B only
    C. Both A and B
    D. Neither A nor B (A.2)

116. While disassembling a manual transmission for an overhaul, all of the following will need to be done EXCEPT:
    A. take and record measurements.
    B. clean all the parts with cleaning solvent.
    C. pay attention to the condition of the parts that are removed.
    D. keep the friction discs in the correct order. (B.5)

117. The speedometer needle bounces when the vehicle is moving. Technician A says that the gear may be worn on the output shaft. Technician B says there could be a bad cable or speedometer head. Who is right?
    A. A only
    B. B only
    C. Both A and B
    D. Neither A nor B (B.13)

118. After inspection of the synchronizer slider for third and fourth gear, the teeth on the slider are found to be rounded and worn. What is the LEAST likely cause of this wear?
    A. third gear
    B. fourth gear
    C. third and fourth gear blocking rings
    D. third and fourth synchronizer hub (C.7)

119. When balancing a driveshaft, all of the following are true EXCEPT:
    A. the driveshaft should be checked for damage before balancing.
    B. screw-type hose clamps can be used for balance weights.
    C. a strobe light is used when checking the balance.
    D. the vehicle suspension should be suspended and not supported. (D.4)

120. A limited slip differential is not working properly. Which of the following would be the LEAST likely cause?
    A. Worn friction plates
    B. Weak spring tension
    C. Stripped teeth on the friction plates
    D. Wrong fluid in the differential (E.3.1)

121. When replacing differential side bearings:
    A. they should be packed with grease.
    B. they are put back on with a hammer and punch.
    C. the bearing races should also be replaced.
    D. the differential case must also be replaced. (C.17)

122. After installing the transmission and cranking the engine, a grinding noise is heard. The LEAST likely cause of this is:
    A. a starter alignment that needs to be adjusted.
    B. missing engine dowels.
    C. a bent or rubbing inspection cover.
    D. a worn transmission main shaft bearing. (A.8)
123. A clicking noise is heard on a front-wheel drive vehicle while turning a corner. The cause of this problem could be a bad:
    A. front axle shaft.
    B. CV inner joint.
    C. torsional damper.
    D. outer CV joint. (D.1)

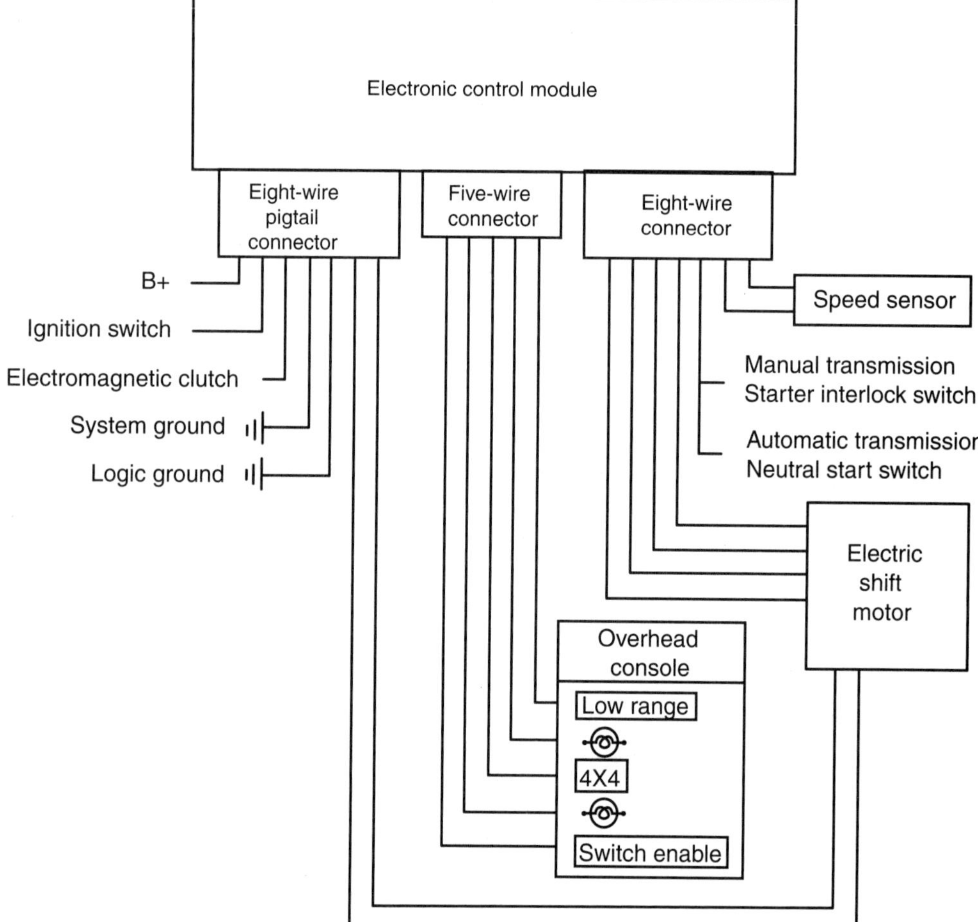

124. Technician A says that all linkages on an electronic-shift transfer case are internal. Technician B says that an open in the shift switch circuit will prevent the transfer case from engaging. Who is right?
    A. A only
    B. B only
    C. Both A or B
    D. Neither A nor B (F.10)

125. A transmission will not engage in reverse when shifted. Which of these is the LEAST likely cause?
A. A bent linkage
B. A broken shift fork
C. A misadjusted linkage
D. Damaged teeth on the gears (B.6)

126. When replacing the ring and pinion gears, all of the following should be replaced EXCEPT the:
A. pinion seal.
B. collapsible pinion spacer.
C. axle seals.
D. spider gears. (E.1.4)

127. After inspecting a pilot bushing, it is found to be worn and must be replaced. Which component is LEAST likely to be checked as a result of the find?
A. The input shaft bearing
B. The input shaft bearing retainer and seal
C. The crankshaft
D. The output shaft bushing (A.6)

128. While discussing clutch assemblies, Technician A says that when replacing a clutch assembly, the flywheel should only be resurfaced if it has excessive runout, scoring, or any other types of imperfections on the face. Technician B says that even if the flywheel looks OK, the flywheel should be removed and resurfaced every time the clutch assembly is replaced. Who is right?
A. A only
B. B only
C. Both A and B
D. Neither A nor B (A.7)

129. Technician A says that when removing a wheel stud you can use a hammer to knock it out. Technician B says you should never use a torch to remove a wheel stud. Who is right?
A. A only
B. B only
C. Both A and B
D. Neither A nor B (E.4.2)

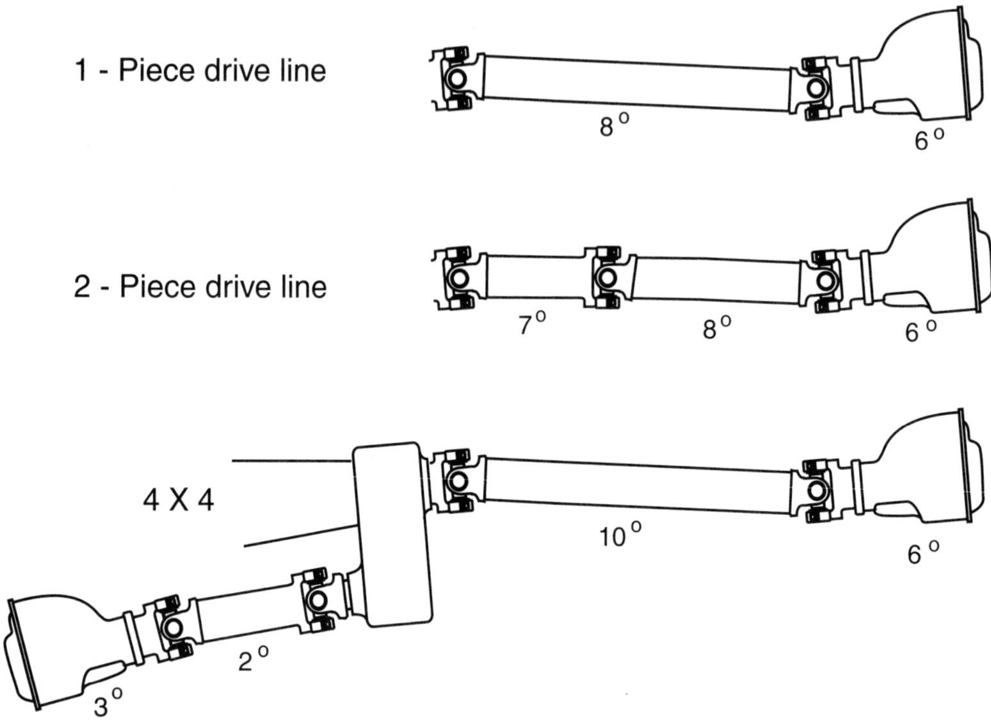

130. If a driveshaft has too great an angle between point A and point B in the figure shown, which components are most likely to be damaged?
   A. The spider gears
   B. The transmission mount
   C. The U-joints
   D. The rear tires (D.6)

131. Technician A says that differential noises can be mistaken for engine noises, wheel bearing noises, or tire noises. Technician B says differential noise will occur when the vehicle is moving. Who is right?
   A. A only
   B. B only
   C. Both A and B
   D. Neither A nor B (C.14)

132. Technician A says that clutch chatter could be caused by an uneven flywheel. Technician B says that oil on the clutch disc can cause clutch chatter. Who is right?
   A. A only
   B. B only
   C. Both A and B
   D. Neither A nor B (A.1)

133. A four-speed transaxle has a clunking noise when driven in first gear and reverse. Technician A says that the gear on the counter shaft could be the problem. Technician B says that the reverse idler gear could be the problem. Who is right?
   A. A only
   B. B only
   C. Both A and B
   D. Neither A nor B (B.1)

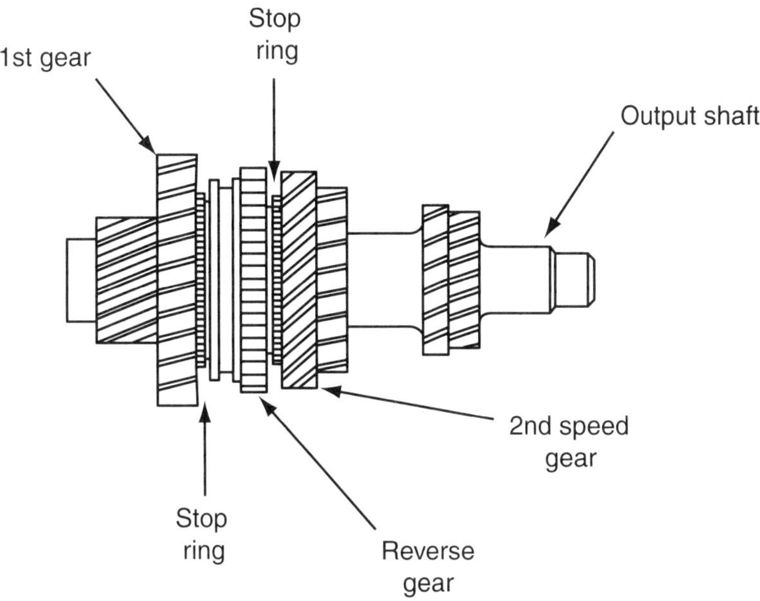

134. While discussing the rebuilding of a manual transaxle, Technician A says that a bent shift fork should be replaced with a new shift fork. Technician B says that the shift fork may be heated and bent back into it's original position. Who is right?
   A. A only
   B. B only
   C. Both A and B
   D. Neither A nor B (C.6)

135. When checking a diagnostic trouble code (DTC) that has to do with the output shaft speed sensor, which of the following should NOT be done next after reading the code?
   A. Install a new sensor.
   B. Inspect the harness for continuity.
   C. Check the service manual.
   D. Inspect the connector. (B.16)

136. Technician A says that inspecting the ring gear tooth pattern is important because it also checks the carrier bearing preload. Technician B says that inspecting the ring gear tooth pattern is important because it also checks the pinion depth. Who is right?
   A. A only
   B. B only
   C. Both A and B
   D. Neither A nor B (E.1.5)

137. A vehicle with a manual transaxle makes noise that is loudest during turns. What would be the most likely cause?
   A. The clutch throwout bearing
   B. Ring gear and pinion teeth
   C. Differential side bearings
   D. The wrong lubricant in the differential (C.1)

138. When replacing a clutch, the pressure plate is found to have small cracks. Technician A says the pressure plate should be replaced. Technician B says the pressure plate should be resurfaced then installed. Who is right?

A. A only

B. B only

C. Both A and B

D. Neither A nor B (A.5)

7 Appendices

Answers to the Test Questions for the Sample Test Section 5

1. A
2. D
3. D
4. D
5. D
6. A
7. C
8. B
9. B
10. A
11. A
12. B
13. C
14. A
15. D
16. B
17. C
18. C
19. C
20. D
21. C
22. A
23. C
24. A
25. B
26. C
27. D
28. C
29. A
30. D
31. A
32. B
33. A
34. D
35. A
36. A
37. A
38. B
39. B
40. A
41. C
42. B
43. D
44. B
45. B
46. C
47. D
48. B
49. A
50. A
51. A
52. C
53. C
54. C
55. B
56. C
57. C
58. C
59. C
60. A
61. C
62. A
63. C
64. C
65. A
66. B
67. A
68. D
69. A
70. D
71. B
72. D
73. D
74. C
75. A
76. D
77. C
78. D
79. A
80. C
81. C
82. B
83. C

Explanations to the Answers for the Sample Test Section 5

Question #1
Answer A is correct. The extension housing seal rides on the drive shaft yoke. A scored yoke could tear the seal.
Answer B is wrong. Excessive output shaft end play would wear the output shaft bushing.
Answer C is wrong. Input shaft play would not affect the extension housing seal.
Answer D is wrong. A output shaft bearing is not in direct contact with the extension housing seal.

Question #2
Answer A is wrong. The drive shaft must be installed in the same position that it was removed from.
Answer B is wrong. The transmission cannot be supported by the input shaft; it could fall and damage the transmission or injure the technician.
Answer C is wrong. The engine support fixture should be installed when the transmission bolts are tight.
Answer D is correct. If the clutch disc is not aligned, the transmission input shaft can be damaged and/or the transmission will not be properly aligned.

Question #3
Answer A is wrong. Technician A is wrong; on a rear-wheel drive vehicle the axle flange is part of the axle shaft and on a front-wheel drive vehicle it is unbolted.
Answer B is wrong. Technician B is wrong; heating the wheel studs can damage the axle flange or the related components.
Answer C is wrong. Both Technicians A and B are wrong.
Answer D is correct. Both Technicians A and B are wrong.

Question #4
Answer A is wrong. Technician A is wrong; a collapsible pinion shaft spacer is not reusable and should be discarded after disassembly.
Answer B is wrong. Technician B is wrong; the ABS toothed ring is not located on the pinion shaft; it is located on the differential case with the ring gear.
Answer C is wrong. Both Technicians A and B are wrong.
Answer D is correct. Both Technicians A and B are wrong.

Question #5
Answer A is wrong. Technician A is wrong; because a dial indicator should be placed near the center of the drive shaft.
Answer B is wrong. Technician B is wrong; a bent drive shaft should be replaced, not repaired.
Answer C is wrong. Both Technicians A and B are wrong.
Answer D is correct. Both Technicians A and B are wrong.

Question #6
Answer A is correct. The linkage alignment pin can only be installed in neutral.
Answer B is wrong. The adjustment must be made with no load on the linkage.
Answer C is wrong. The adjustment must be made with no load on the linkage.
Answer D is wrong. The adjustment must be made with no load on the linkage.

Question #7
Answer A is wrong. Friction plate thickness is measured with a micrometer.
Answer B is wrong. Steel plate thickness is measured with a micrometer.
Answer C is correct. A feeler gauge can be used to determine the required shim thickness.
Answer D is wrong. Spring tension is not measured with a feeler gauge.

Question #8
Answer A is wrong. The tips should be sharp and pointed.
Answer B is correct.
Answer C is wrong. There must be specified clearance between the blocker rings and the gear teeth.
Answer D is wrong. The synchronizer sleeve must slide freely on the hub.

Question #9
Answer A is wrong. If the synchronizer sleeve does not slide smoothly over the blocker ring and gear teeth, shifting will not be smooth, but it will not be prevented.
Answer B is correct.
Answer C is wrong. Only Technician B is right.
Answer D is wrong. Technician B is right.

Question #10
Answer A is correct. If the differential case runout is excessive the ring gear runout will be excessive.
Answer B is wrong. Side bearing preload does not affect ring gear runout.
Answer C is wrong. Ring gear runout has no relation to the side gear end play.
Answer D is wrong. Once the ring gear bolts are torqued they do not affect ring gear runout.

Question #11
Answer A is correct.
Answer B is wrong. A wheel bearing will not make a clunking noise when it is damaged.
Answer C is wrong. Only Technician A is right.
Answer D is wrong. Technician A is right.

Question #12
Answer A is wrong. The differential bearing preload does not affect axle flange runout.
Answer B is correct. A bent axle shaft will cause excessive flange runout.
Answer C is wrong. Pinion depth has nothing to do with axle flange runout.
Answer D is wrong. Limited slip clutch discs have nothing to do with axle flange runout.

Question #13
Answer A is wrong. The pinion is marked with +3; therefore, add three thousandths to the shim that provides a slight drag between the gauge block and the gauge tube.
Answer B is wrong. The technician does not want to add a shim that is smaller than the measured clearance.
Answer C is correct. The pinion is marked with +3, therefore add three thousandths to 0.063 inch (1.6 mm).
Answer D is wrong. Adding this shim would be more than the measured clearance and the pinion marking added together.

Question #14
Answer A is correct.
Answer B is wrong. A driveshaft vibration will not have a constant speed since the drive shaft speed is always changing with incorrect U-joint angles.
Answer C is wrong. Only Technician A is right.
Answer D is wrong. Technician A is right.

Question #15
Answer A is wrong. The reverse idler gear only rotates in reverse.
Answer B is wrong. The reverse idler gear only rotates in reverse.
Answer C is wrong. The reverse idler gear only rotates in reverse.
Answer D is correct.

Question #16
Answer A is wrong. If the vehicle's U-joints were worn to the point of causing a vibration, it would not only be evident while cornering, but at all speeds.
Answer B is correct. Outboard U-joint vibration increases during a turn.
Answer C is wrong. Only Technician B is right.
Answer D is wrong. Technician B is right.

Question #17
Answer A is wrong. Both Technicians A and B are right because with the engine at idle and the transmission in neutral with the clutch released, the countershaft will spin. This will create a growling from the countershaft and bearings that are pitted and scored.
Answer B is wrong. Both Technicians A and B are right because the countershaft spins in all gears and will create a growling from the countershaft and bearings that are pitted and scored.
Answer C is correct. Both Technicians A and B are right.
Answer D is wrong. Both Technicians A and B are right.

Question #18
Answer A is wrong. Tightening both adjusters would make the backlash tighter.
Answer B is wrong. Both adjusters must be changed at the same time.
Answer C is correct. This adjustment would increase the backlash.
Answer D is wrong. This adjustment would also decrease the backlash.

Question #19
Answer A is wrong. A worn blocker ring would only cause a noise during shifting.
Answer B is wrong. A worn first-gear synchronizer ring would only cause a noise during shifting.
Answer C is correct. A chipped and worn first-speed gear would produce a growl in first gear only.
Answer D is wrong. A worn main shaft bearing would cause noise in all gears.

Question #20
Answer A is wrong. Technician A is wrong; the drive pinion should be moved away from the drive gear.
Answer B is wrong. Technician B is wrong; the drive pinion should be moved away from the drive gear.
Answer C is wrong. Both Technicians A and B are wrong.
Answer D is correct. Both Technicians A and B are wrong.

Question #21
Answer A is wrong. Both Technicians A and B are right because if the front suspension is not aligned, the drive axles will not be properly aligned.
Answer B is wrong. Both Technicians A and B are right because the lower control arms are connected to the cradle; and if the cradle is not properly aligned, the front suspension will not be properly aligned.
Answer C is correct. Both Technicians A and B are right.
Answer D is wrong. Both Technicians A and B are right.

Question #22
Answer A is correct. A plugged vent can cause enough internal pressure to force fluid past a seal
Answer B is wrong. An outer drive axle joint does not come into contact with the drive seal.
Answer C is wrong. Only Technician A is right.
Answer D is wrong. Technician A is right.

Question #23
Answer A is wrong. First gear is at the rear of the transmission, away from the input shaft.
Answer B is wrong. Second gear is the second closest gear to the rear of the transmission.
Answer C is correct. Fourth gear is the closest to the input shaft.
Answer D is wrong. Reverse is at the rear of the transmission.

Question #24
Answer A is correct. A loose pinion nut reduces pinion preload and lets the pinion clunk on the ring gear.
Answer B is wrong. Only a pinion nut that is overtorqued will make a growling noise when the vehicle is in motion.
Answer C is wrong. Only Technician A is right.
Answer D is wrong. Technician A is right.

Question #25
Answer A is wrong. While driving straight ahead, the side gears both turn at the same speed; they would not be the cause of the noise.
Answer B is correct. Insufficient backlash can cause ring-and-pinion whine.
Answer C is wrong. Only Technician B is right.
Answer D is wrong. Technician B is right.

Question #26
Answer A is wrong. Technician B is also right; a broken oil feeder will not get lubricant to the output shaft and bearings.
Answer B is wrong. Technician A is also right; only the correct lubricant will flow through the system as designed and perform as intended.
Answer C is correct. Both Technicians A and B are right.
Answer D is wrong. Both Technicians A and B are right.

Question #27
Answer A is wrong. Technician A is wrong; transaxles do not use threaded adjusters to adjust the side bearing preload.
Answer B is wrong. Technician B is wrong; shims are installed as needed to adjust the side bearing preload.
Answer C is wrong. Both Technicians A and B are wrong.
Answer D is correct. Both Technicians A and B are wrong.

Question #28
Answer A is wrong. Speedometer drive gears are never located on the input shaft.
Answer B is wrong. The transfer gear does not rotate at the same rpm as the drive axle.
Answer C is correct.
Answer D is wrong. The drive axle inner hub is not located in the transaxle.

Question #29
Answer A is correct. A broken shift fork would not usually cause noise.
Answer B is wrong. If the reverse idler gear teeth were damaged, there would be a noise.
Answer C is wrong. Only Technician A is right.
Answer D is wrong. Technician A is right.

Question #30
Answer A is wrong. Technician A is wrong; if the synchronizer is reversed, it would not function.
Answer B is wrong. Technician B is wrong; synchronizer sleeves are machined to fit in only one way. If reversed, it would not fit together.
Answer C is wrong. Both Technicians A and B are wrong.
Answer D is correct. Both Technicians A and B are wrong.

Question #31
Answer A is correct.
Answer B is wrong. After the transmission was shifted into third gear, there would be no vibration.
Answer C is wrong. Worn dog teeth on the blocker ring would cause shifting problems only.
Answer D is wrong. Only shifting problems would occur.

Question #32
Answer A is wrong. The bearings and the surfaces that they ride on can be damaged if operated without lubricant.
Answer B is correct.
Answer C is wrong. Only Technician B is right.
Answer D is wrong. Technician B is right.

Question #33
Answer A is correct. A worn adapter bushing would allow the speedometer pinion to move excessively.
Answer B is wrong. A worn extension housing bushing would only cause an audible thunk when the vehicle was put into gear or when accelerated.
Answer C is wrong. Main shaft end play would not affect the output shaft which is where the speedometer drive is attached.
Answer D is wrong. The only electrical sensor that would cause an erratic speedometer reading would be the vehicle speed sensor.

Question #34
Answer A is wrong. When removing the drive shaft, only two marks need to be applied; one on the drive shaft and the other on the pinion flange.
Answer B is wrong. When replacing the U-joints, the drive shaft does not need to be marked.
Answer C is wrong. Drive shaft runout measurement is performed with a dial indicator.
Answer D is correct. Drive shaft balance is checked with four marks 90 degrees apart.

Question #35
Answer A is correct.
Answer B is wrong. The thicker the gear oil, the thicker the viscosity.
Answer C is wrong. Only Technician A is right.
Answer D is wrong. Technician A is right.

Question #36
Answer A is correct. Removing shims from the right and adding an equal number to the left moves the ring gear inward to decrease backlash without changing preload.
Answer B is wrong. Adding shims equally to both sides will change preload without affecting backlash.
Answer C is wrong. Removing shims from the left and adding an equal number to the right moves the ring gear outward to increase backlash.
Answer D is wrong. Removing shims equally from both sides will change preload without affecting backlash.

Question #37
Answer A is correct.
Answer B is wrong. Loctite® is a thread sealer.
Answer C is wrong. Only Technician A is right.
Answer D is wrong. Technician A is right.

Question #38
Answer A is wrong. The friction plates are splined to the side gear which is splined to the axle shaft.
Answer B is correct. Only the friction plates are splined to the axle shafts.
Answer C is wrong. The preload spring applies pressure to the clutch packs.
Answer D is wrong. Special lubricant is required.

Question #39
Answer A is wrong. A bent shift rail usually will not cause the transmission to jump out of gear.
Answer B is correct. A bent shift rail usually causes hard shifting.
Answer C is wrong. A bent shift rail will not usually cause gear clash during shifting.
Answer D is wrong. A bent shift rail will not cause gear noise.

Question #40
Answer A is correct. If the shaft has damage in the seal area, it should be replaced.
Answer B is wrong. It is always a good practice to replace the axle seals while the axle shafts are removed.
Answer C is wrong. Always stand the axle shaft on its end so that the splines do not contact the floor which could cause damage.
Answer D is wrong. Some require the use of a slide hammer to remove them from the differential.

Question #41
Answer A is wrong. If a pilot bearing is damaged, it will make more noise with the clutch disengaged.
Answer B is wrong. If the input shaft pilot bearing contact surface is damaged, it will make more noise with the clutch disengaged.
Answer C is correct. The input is not turning with the clutch depressed, but is turning with the clutch engaged.
Answer D is wrong. A mainshaft bearing failure would make noise no matter what position the clutch was in.

Question #42
Answer A is wrong. A hydraulic clutch system does not require free play.
Answer B is correct. Air in the system will prevent the clutch from disengaging properly.
Answer C is wrong. If the clutch facings were worn, it would have no effect on clutch disengagement.
Answer D is wrong. A scored pressure plate would not cause clutch disc disengagement problems.

Question #43
Answer A is wrong. The dog teeth only would cause noise when shifting into second gear.
Answer B is wrong. A vibration in second gear would be an indication of gear problems.
Answer C is wrong. Difficult shifting would happen in second gear only.
Answer D is correct.

Question #44
Answer A is wrong. A worn fourth-gear synchronizer would only affect shifting into fourth gear.
Answer B is correct. A sticking clutch disc would affect all gears.
Answer C is wrong. End play would not normally cause gear clash.
Answer D is wrong. If only the 3-4 shifter fork was worn, only shifting into third or fourth gear would cause gear clash.

Question #45
Answer A is wrong. Drive pinion bearings would not cause a vibration while turning. They would cause noise all the time.
Answer B is correct. Differential side bearings are under load only during a turn.
Answer C is wrong. Only Technician B is right.
Answer D is wrong. Technician B is right.

Question #46
Answer A is wrong. Technician B is also right; the switch closes when the transmission is shifted into reverse and usually receives battery voltage with the ignition in RUN.
Answer B is wrong. Technician A is also right; the backup lamp switch is normally open and closes when the transmission is placed in reverse.
Answer C is correct. Both Technicians A and B are right.
Answer D is wrong. Both Technicians A and B are right.

Question #47
Answer A is wrong. The ring gear runout should be checked to determine the condition of the trueness of the ring gear.
Answer B is wrong. The case side play should be measured before disassembly to determine the condition of the case.
Answer C is wrong. Bearing caps should be marked and reinstalled in the same position that they were installed.
Answer D is correct. The side bearings should be lubricated before installation.

Question #48
Answer A is wrong. A worn speedometer drive and driven gears would not cause premature housing wear.
Answer B is correct. Uneven mating surfaces can offset the extension housing and cause uneven bushing wear.
Answer C is wrong. A plugged transmission vent opening would cause the fluid to overheat and overflow.
Answer D is wrong. Main shaft end play does not usually cause bushing wear.

Question #49
Answer A is correct. If the ring and pinion were damaged, the vehicle would have problems moving.
Answer B is wrong. Differential gears do not affect straight-ahead movement.
Answer C is wrong. Only Technician A is right.
Answer D is wrong. Technician A is right.

Question #50
Answer A is correct.
Answer B is wrong. A worn release bearing makes more noise when the clutch pedal is depressed and the release bearing is in contact with the pressure plate.
Answer C is wrong. Only Technician A is right.
Answer D is wrong. Technician A is right.

Question #51
Answer A is correct. The dial indicator measures crankshaft movement forward and backward in the block.
Answer B is wrong. Crankshaft warpage can only be checked with the crankshaft out of the engine.
Answer C is wrong. Main bearing wear has to be measured after the engine is disassembled.
Answer D is wrong. The bell housing has to be installed on the engine block to measure the engine block alignment.

Question #52
Answer A is wrong. Technician A is right that you can test specified resistance of a pickup coil VSS with an ohmmeter.
Answer B is wrong. Technician B is right that you can check the VSS output signal with an oscilloscope.
Answer C is correct. Both Technicians A and B are right.
Answer D is wrong. Both Technicians A and B are right.

Question #53
Answer A is wrong. Technician B is also right; if the vacuum motor does not operate, the 4WD system will not engage.
Answer B is wrong. Technician A is also right; the vacuum motor needs a good vacuum source to operate.
Answer C is correct.
Answer D is wrong. Both Technicians A and B are right.

Question #54
Answer A is wrong. Both Technicians A and B are right because if the linkage is not adjusted correctly, it may not shift into gear.
Answer B is wrong. Both Technicians A and B are right because if the linkage is not adjusted correctly, it may be hard to shift out of gear.
Answer C is correct. Both Technicians A and B are right.
Answer D is wrong. Both Technicians A and B are right.

Question #55
Answer A is wrong. This choice has nothing to do with the measurement shown.
Answer B is correct.
Answer C is wrong. The differential shim should not be the same as the end play measurement.
Answer D is wrong. The transaxle bolts should be tightened to the specified torque.

Question #56
Answer A is wrong. Most inner tripods are held in place with snap rings.
Answer B is wrong. Special pliers are used to crimp the outer boot clamps.
Answer C is correct. All of the supplied grease should be used.
Answer D is wrong. Worn CV joints can make a clicking noise when cornering because of lack of lubricant.

Question #57
Answer A is wrong. Technician B is also right; the center support bearing will get louder as the vehicle accelerates or decelerates.
Answer B is wrong. Technician A is also right; a damaged outer rear axle bearing will be louder as the vehicle accelerates or decelerates than it will be at steady cruising speed.
Answer C is correct. Both Technicians A and B are right.
Answer D is wrong. Both Technicians A and B are right.

Question #58
Answer A is wrong. Packing a bearing before installation is to pack the bearing with grease. It does not affect preload.
Answer B is wrong. Bearing preload is the amount of pressure applied to a bearing upon assembly of the transmission. It does not change during operation.
Answer C is correct.
Answer D is wrong. Bearing preload is adjustable.

Question #59
Answer A is wrong. Technician B is also right; if the vent is left on top of the transfer case, water may enter through the vent.
Answer B is wrong. Technician A is also right; as the transfer case warms up while the vehicle is being driven, pressure builds up in the case which can force lubricant out at the weakest point in a seal or gasket.
Answer C is correct. Both Technicians A and B are right.
Answer D is wrong. Both Technicians A and B are right.

Question #60
Answer A is correct.
Answer B is wrong. The annulus gear is locked to the case so that it cannot rotate.
Answer C is wrong. Only Technician A is right.
Answer D is wrong. Technician A is right.

Question #61
Answer A is wrong. Technician B is also right; the detent springs keep the shift rails from moving.
Answer B is wrong. Technician A is also right; the synchronizer eliminates gear clash when shifting and keeps a transmission in gear.
Answer C is correct. Both Technicians A and B are right.
Answer D is wrong. Both Technicians A and B are right.

Question #62
Answer A is correct.
Answer B is wrong. Not all transfer cases are made of cast iron. It depends on the application whether the transfer case is removed with the transmission or by itself.
Answer C is wrong. Only Technician A is right.
Answer D is wrong. Technician A is right.

Question #63
Answer B is wrong. Technician A is also right; the plates require a special lubricant to keep from chattering.
Answer A is wrong. Technician B is also right; the friction plates wear and can develop hotspots with use.
Answer C is correct. Both Technicians A and B are right.
Answer D is wrong. Both Technicians A and B are right.

Question #64
Answer A is wrong. Technician B is also right; the clutch and pressure plate will be sitting closer to the engine and this may be too far for the slave cylinder to operate.
Answer B is wrong. Technician A is also right; material is removed from the flywheel when resurfaced, and this would allow the clutch disc to sit closer to the flywheel bolts.
Answer C is correct. Both Technicians A and B are right.
Answer D is wrong. Both Technicians A and B are right.

Question #65
Answer A is correct. This is the only choice in which the input shaft and pilot bearing are turning at different speeds to cause a rattle.
Answer B is wrong. While decelerating in high gear, the load on the pilot bearing is not at its greatest.
Answer C is wrong. While accelerating in low gear, the load on the pilot bearing is not at its greatest.
Answer D is wrong. There is almost no load on the pilot bearing while the engine is in neutral and the clutch engaged.

Question #66
Answer A is wrong. The squeaking noise will increase as the vehicle speed increases.
Answer B is correct.
Answer C is wrong. Only Technician B is right.
Answer D is wrong. Technician B is right.

Question #67
Answer A is correct.
Answer B is wrong. An axle shaft with excessive runout will cause vibration at all times no matter what the vehicle speed.
Answer C is wrong. Only Technician A is right.
Answer D is wrong. Technician A is right.

Question #68
Answer A is wrong. Worn U-joints can cause driveline vibrations.
Answer B is wrong. Worn front-drive axle joints will cause vibrations when turning corners.
Answer C is wrong. Incorrect drive shaft angles will cause vibrations and premature wear of the U-joints.
Answer D is correct. Slip joint spline wear is uncommon and unlikely to cause noticeable vibration.

Question #69
Answer A is correct. Bearings must always be lubricated for this adjustment.
Answer B is wrong. If the pinion nut is overtorqued, the assembly has to be disassembled and a new crush sleeve has to be installed before attempting to obtain the proper turning torque.
Answer C is wrong. Only Technician A is right.
Answer D is wrong. Technician A is right.

Question #70
Answer A is wrong. A worn clutch release bearing makes a squeaking or rattling noise when the clutch pedal is depressed.
Answer B is wrong. A worn pilot bearing would not likely cause clutch chatter, but would cause shifting problems.
Answer C is wrong. Excessive input shaft play would cause noises and shifting problems, but not likely clutch chatter.
Answer D is correct. The torsional springs are designed to help reduce clutch chatter.

Question #71
Answer A is wrong. Scored side bearings would give inaccurate readings while measuring case runout.
Answer B is correct. Case runout is part of total ring gear runout.
Answer C is wrong. Only Technician B is right.
Answer D is wrong. Technician B is right.

Question #72
Answer A is wrong. Hard shifting would only be caused if the clutch pedal free play was excessive.
Answer B is wrong. Incomplete clutch release would be caused by excessive pedal free play.
Answer C is wrong. Improper clutch pedal free play cannot cause transaxle gear damage.
Answer D is correct. Lack of free play may keep the clutch from fully engaging.

Question #73
Answer A is wrong. Misalignment would not affect the free play but it would cause premature throwout bearing wear.
Answer B is wrong. A growling noise when the clutch pedal is depressed would indicate that the throwout bearing is worn.
Answer C is wrong. After the clutch is fully engaged, there would not be any vibration.
Answer D is correct. Misalignment would allow one part of the clutch to grab sooner than the rest of the clutch disc.

Question #74
Answer A is wrong. Technician B is also right; the wheel will vibrate more as the vehicle accelerates.
Answer B is wrong. Technician A is also right; a drive shaft will vibrate more as the vehicle accelerates.
Answer C is correct. Both Technicians A and B are right.
Answer D is wrong. Both Technicians A and B are right.

Question #75
Answer A is correct. Detent springs help to hold the shift rail in the engaged position.
Answer B is wrong. If the fourth gear teeth were worn, the transaxle would jump out of fourth gear.
Answer C is wrong. Only Technician A is right.
Answer D is wrong. Technician A is right.

Question #76
Answer A is wrong. Over time, the transfer case chain stretches.
Answer B is wrong. Select-fit thrust washers should be measured and used for assembly of the transfer case.
Answer C is wrong. During disassembly, the transfer case components should be thoroughly inspected.
Answer D is correct. All the transfer case parts should be cleaned then lubricated before assembly.

Question #77
Answer A is wrong. The washer thickness has to be calculated individually for each side.
Answer B is wrong. The side gear end play cannot be measured with the shims behind the gears.
Answer C is correct.
Answer D is wrong. The purpose is to take up excess clearance, not to provide preload.

Question #78
Answer A is wrong. Both Technicians A and B are wrong.
Answer B is wrong. Both Technicians A and B are wrong.
Answer C is wrong. Both Technicians A and B are wrong.
Answer D is correct. Automatic locking hubs are lubricated with oil, not packed with grease.

Question #79
Answer A is correct.
Answer B is wrong. Overheating the clutch facing cannot affect bell housing runout.
Answer C is wrong. Only Technician A is right.
Answer D is wrong. Technician A is right.

Question #80
Answer A is wrong. Excessive crankshaft end play does not cause clutch chatter.
Answer B is wrong. Loose main engine bearings do not cause clutch chatter.
Answer C is correct. A scored pressure plate will not apply an even force to the clutch disc.
Answer D is wrong. Improper pressure plate-to-flywheel position would not necessarily cause clutch chatter.

Question #81
Answer A is wrong. Technician B is also right; too much lubricant may leak from the axle seals.
Answer B is wrong. Technician A is also right; the axle vent will allow air to escape from the axle and not have any pressure buildup.
Answer C is correct. Both Technicians A and B are right.
Answer D is wrong. Both Technicians A and B are right.

Question #82
Answer A is wrong. A dial indicator is used to measure differential case runout.
Answer B is correct.
Answer C is wrong. A dial indicator is used to measure differential case runout.
Answer D is wrong. A dial indicator is used to measure differential case runout.

Question #83
Answer A is wrong. Technician B is also right; a bearing and race are a matched set—a new bearing should never be used with an old race.
Answer B is wrong. Technician A is also right; if the bearings are not damaged, they may be reused.
Answer C is correct. Both Technicians A and B are right.
Answer D is wrong. Both Technicians A and B are right.

Answers to the Test Questions for the Additional Test Questions Section 6

1.	C	36.	A	71.	A	106.	D
2.	B	37.	B	72.	A	107.	A
3.	C	38.	A	73.	D	108.	D
4.	D	39.	D	74.	C	109.	A
5.	A	40.	B	75.	B	110.	D
6.	A	41.	A	76.	B	111.	A
7.	A	42.	B	77.	C	112.	D
8.	B	43.	B	78.	B	113.	D
9.	C	44.	C	79.	B	114.	A
10.	C	45.	C	80.	D	115.	C
11.	D	46.	C	81.	A	116.	D
12.	A	47.	B	82.	C	117.	C
13.	C	48.	D	83.	D	118.	D
14.	B	49.	A	84.	C	119.	D
15.	C	50.	C	85.	A	120.	C
16.	C	51.	B	86.	D	121.	C
17.	D	52.	C	87.	A	122.	D
18.	C	53.	D	88.	B	123.	D
19.	C	54.	C	89.	C	124.	C
20.	D	55.	A	90.	C	125.	D
21.	A	56.	A	91.	A	126.	D
22.	D	57.	C	92.	C	127.	D
23.	B	58.	C	93.	C	128.	B
24.	A	59.	C	94.	A	129.	C
25.	A	60.	C	95.	A	130.	C
26.	C	61.	C	96.	B	131.	C
27.	D	62.	B	97.	C	132.	C
28.	C	63.	A	98.	A	133.	A
29.	A	64.	A	99.	D	134.	A
30.	D	65.	B	100.	B	135.	A
31.	A	66.	C	101.	C	136.	B
32.	D	67.	D	102.	B	137.	C
33.	D	68.	D	103.	B	138.	A
34.	A	69.	C	104.	D		
35.	D	70.	D	105.	C		

Explanations to the Answers for the Additional Test Questions Section 6

Question #1
Answer A is wrong. Technician B is also right; the floor- and column-mounted linkages can be adjusted.
Answer B is wrong. Technician A is also right; there are no internal adjustments. (The floor and column-mounted linkages can be adjusted, however.)
Answer C is correct. Both Technicians A and B are right.
Answer D is wrong. Both Technicians A and B are right.

Question #2
Answer A is wrong. Replacing the engine mounts would not affect bell housing bore alignment.
Answer B is correct.
Answer C is wrong. Bell housing shims would not correct a bell housing bore alignment condition.
Answer D is wrong. The clutch disc does control bell housing bore alignment.

Question #3
Answer A is wrong. The differential cover may be left on while measuring axle shaft end play.
Answer B is wrong. The vehicle does not have to be in neutral to measure axle shaft end play.
Answer C is correct. A dial indicator will measure axle shaft end play at the axle flange.
Answer D is wrong. Axle shaft end play cannot be measured with the axle shaft removed from the axle housing.

Question #4
Answer A is wrong. A plugged axle housing vent may cause excessive pressure that will lead to a leak.
Answer B is wrong. If the differential is overfilled, the fluid may leak past a housing seal.
Answer C is wrong. Worn axle shaft bearings will put a load on the axle shaft seal, and could cause a leak.
Answer D is correct.

Question #5
Answer A is correct. When the clutch is disengaged, the pilot bearing rotates on the input shaft, which can let the bearing make noise
Answer B is wrong. If the clutch is disengaged, the input shaft on the transmission is not spinning, eliminating the input shaft as a source.
Answer C is wrong. Only Technician A is right.
Answer D is wrong. Technician A is right.

Question #6
Answer A is correct.
Answer B is wrong. The reverse idler gear rides on needle bearings and is not splined in the center bore.
Answer C is wrong. Only Technician A is right.
Answer D is wrong. Technician A is right.

Question #7
Answer A is correct.
Answer B is wrong. The gasket or sealer will seal the mating surfaces if the burrs have been filed off.
Answer C is wrong. Only Technician A is right.
Answer D is wrong. Technician A is right.

Question #8
Answer A is wrong. The flywheel has no fixed relationship to the transmission input shaft.
Answer B is correct. The flywheel has an important relationship to the crankshaft because it is a major factor in engine balance.
Answer C is wrong. The flywheel has no fixed relationship to the clutch disc.
Answer D is wrong. The flywheel has no fixed relationship to the bell housing.

Question #9
Answer A is wrong. The transmission should be in neutral when checking driveshaft runout.
Answer B is wrong. A magnetic-base dial indicator should be installed near the center of the driveshaft.
Answer C is correct. This is not an acceptable way of inspecting a driveshaft.
Answer D is wrong. The driveshaft surface must be free of debris to ensure an accurate reading.

Question #10
Answer A is wrong. The clutch pedal will not have any free play in a self-adjusting system.
Answer B is wrong. The clutch pedal will not have any free play in a self-adjusting system.
Answer C is correct. Self-adjusting clutches do not normally have any free play, which causes the release bearing to contact the clutch release mechanism continuously.
Answer D is wrong. A clutch with a self-adjusting cable does not have an overcenter assist spring.

Question #11
Answer A is wrong. In any position other than choice D, centrifugal force will try to throw grease out of the fitting.
Answer B is wrong. In any position other than choice D, centrifugal force will try to throw grease out of the fitting.
Answer C is wrong. In any position other than choice D, centrifugal force will try to throw grease out of the fitting.
Answer D is correct. This is the only position in which centrifugal force will not try to throw grease out of the fitting.

Question #12
Answer A is correct. The engine will need to be supported without the transaxle.
Answer B is wrong. The engine oil does not need to be drained when removing the transaxle.
Answer C is wrong. The negative battery cable should be disconnected; not the positive.
Answer D is wrong. The engine does not need to be removed when removing the transaxle from many vehicles.

Question #13
Answer A is wrong. The U-joints are not in the transfer case.
Answer B is wrong. The U-joints are not in the transfer case.
Answer C is correct. The U-joints are outside the transfer case.
Answer D is wrong. A transfer case output bearing can cause transfer case noise.

Question #14
Answer A is wrong. The remote vent tube on a differential does not increase pressure.
Answer B is correct. The remote vent tube is located above a point where moisture could enter.
Answer C is wrong. The remote vent tube on a differential does not keep lubricant from coming out of the differential.
Answer D is wrong. The remote vent tube on a differential is not used to add lubricant to the differential.

Question #15
Answer A is wrong. The transfer case usually will shift without fluid.
Answer B is wrong. The transfer case will shift into 4WD but will not be able to put power to the front differential.
Answer C is correct.
Answer D is wrong. A manual shift transfer case does not have an electronic shift motor.

Question #16
Answer A is wrong. The axle shafts must be removed from the transaxle to remove the transaxle differential.
Answer B is wrong. The transaxle must be removed from the vehicle and the case opened to remove the transaxle differential.
Answer C is correct.
Answer D is wrong. The lower control arms are not related to transaxle removal.

Question #17
Answer A is wrong. This could have a direct effect on the seal and sealing surface.
Answer B is wrong. This could have a direct effect on the seal and sealing surface.
Answer C is wrong. This could have a direct effect on the seal and sealing surface.
Answer D is correct. The brake drum does not have a direct effect on the seal and sealing surface.

Question #18
Answer A is wrong. A sign of a bad outer CV joint is a clunking or clicking while turning.
Answer B is wrong. A bad CV joint could also make noise while the vehicle is going straight if it is excessively worn.
Answer C is correct. Both Technicians A and B are right.
Answer D is wrong. Both Technicians A and B are right.

Question #19
Answer A is wrong. The axle seals are not designed to hold back large quantities of axle fluid.
Answer B is wrong. A bad axle bearing would allow the axle to apply pressure to the seal.
Answer C is correct. Both Technicians A and B are right.
Answer D is wrong. Both Technicians A and B are right.

Question #20
Answer A is wrong. A broken detent spring could cause the transmission to jump out of gear.
Answer B is wrong. A broken detent spring could cause harsh shifting.
Answer C is wrong. A broken detent spring could cause the transmission to lock up between gears.
Answer D is correct. A faulty bearing or bushing will cause a growling noise.

Question #21
Answer A is correct.
Answer B is wrong. Bushings would cause too much friction, resulting in worn bushings.
Answer C is wrong. Only Technician A is right.
Answer D is wrong. Technician A is right.

Question #22
Answer A is wrong. The bearings do not need to be replaced to correct the pinion gear rotating torque.
Answer B is wrong. The pinion gear cannot be replaced without replacing the ring gear.
Answer C is wrong. The pinion gear nut should never be loosened to compensate for incorrect pinion gear rotating torque.
Answer D is correct.

Question #23
Answer A is wrong. Only worn bushings must be replaced.
Answer B is correct.
Answer C is wrong. Only Technician B is right.
Answer D is wrong. Technician B is right.

Question #24
Answer A is correct. The input bearing retainer and seal work in conjunction with the input bearing.
Answer B is wrong. The counter shaft bearing is not the most likely component to be damaged.
Answer C is wrong. Second gear is not the most likely component to be damaged.
Answer D is wrong. The counter gear is not the most likely component to be damaged.

Question #25
Answer A is correct.
Answer B is wrong. The transaxle case should be split after it is removed from the vehicle.
Answer C is wrong. The engine should not need to be removed to remove the transaxle.
Answer D is wrong. Only part of the transaxle needs to be disassembled to remove the differential assembly.

Question #26
Answer A is wrong. When replacing a speedometer cable core, the old gear can be used.
Answer B is wrong. A different gear may have a different number of teeth, which would change the speedometer reading.
Answer C is correct. Both Technicians A and B are right.
Answer D is wrong. Both Technicians A and B are right.

Question #27
Answer A is wrong. The extension housing bushing should be inspected.
Answer B is wrong. The drive shaft yoke should be inspected.
Answer C is wrong. The extension housing gasket should be inspected.
Answer D is correct. The input shift bearing has no effect on the extension housing seal.

Question #28
Answer A is wrong. A new bushing should be lubricated with motor oil where the transmission input shaft is installed.
Answer B is wrong. A new bearing should be lubricated with grease where the transmission input shaft is installed.
Answer C is correct.
Answer D is wrong. Both Technicians A and B are right.

Question #29
Answer A is correct. Engine oil can contaminate the clutch disc.
Answer B is wrong. The engine should always be supported when the transmission is removed.
Answer C is wrong. Only Technician A is right.
Answer D is wrong. Technician A is right.

Question #30
Answer A is wrong. Worn axle shaft bearings may cause axle shaft seal leakage.
Answer B is wrong. A plugged axle vent may cause axle shaft seal leakage.
Answer C is wrong. Scored axle shafts may damage the seal, which could lead to a leak.
Answer D is correct. The seal is made of heat-treated rubber to resist heat damage.

Question #31
Answer A is correct. Sharp edges help to engage the synchronizer.
Answer B is wrong. The sharp edges will not prevent the transmission from jumping out of gear.
Answer C is wrong. Only Technician A is right.
Answer D is wrong. Technician A is right.

Question #32
Answer A is wrong. Excessive end play in the counter gear could cause transmission case damage.
Answer B is wrong. Excessive end play in the counter gear could cause gear teeth damage.
Answer C is wrong. Excessive end play in the counter gear could cause counter shaft damage.
Answer D is correct.

Question #33
Answer A is wrong. The universal joint should not be greased so much that grease squirts out of the caps.
Answer B is wrong. Drive shaft balance is not affected by the amount of grease in the universal joint.
Answer C is wrong. Both Technicians A and B are wrong.
Answer D is correct. Both Technicians A and B are wrong.

Question #34
Answer A is correct. Hunting-type gearsets do not require the gearset to be timed.
Answer B is wrong. Loose ring gear bolts may cause a gear chuckle or knocking while driving.
Answer C is wrong. Damaged ring gear and pinion gear teeth may cause a ticking noise while driving.
Answer D is wrong. On some ring gearsets, the painted teeth on the ring gear must be matched to the painted teeth on the pinion gear.

Question #35
Answer A is wrong. A short in the backup light switch may cause this problem.
Answer B is wrong. A misadjusted linkage could cause this problem.
Answer C is wrong. A short in the wiring harness may cause this problem.
Answer D is correct. The brake light switch has no effect on the backup lights.

Question #36
Answer A is correct.
Answer B is wrong. The release bearings move toward the pressure plate to disengage the clutch.
Answer C is wrong. Only Technician A is right.
Answer D is wrong. Technician A is right.

Question #37
Answer A is wrong. This will not affect flywheel runout.
Answer B is correct. Crankshaft end play can influence flywheel runout.
Answer C is wrong. This will not affect flywheel runout.
Answer D is wrong. This will not affect flywheel runout.

Question #38
Answer A is correct. Incorrect preload and backlash cause incorrect gear tooth contact, which usually produces noise.
Answer B is wrong. Using the wrong differential lube will not cause a constant whining noise.
Answer C is wrong. If the side gears were damaged, the noise would only occur when turning.
Answer D is wrong. If the spider gears were damaged, the noise would only occur when turning.

Question #39
Answer A is wrong. A worn second gear may cause the transaxle to jump out of second gear.
Answer B is wrong. Excessive main shaft end play will cause the transaxle to jump out of second gear.
Answer C is wrong. A shifter linkage out of adjustment may cause the transaxle to jump out of second gear.
Answer D is correct. A worn ring gear in the differential will not cause the transaxle to jump out of second gear.

Question #40
Answer A is wrong. If the clearance on the fourth-speed gear is less than specified, it would only affect the transition into fourth gear. It could not cause noise when fourth gear is engaged.
Answer B is correct.
Answer C is wrong. Only Technician B is right.
Answer D is wrong. Technician B is right.

Question #41
Answer A is correct. The gear is machined into the output shaft.
Answer B is wrong. The drive gear is made out of nylon.
Answer C is wrong. The gear is machined into the output shaft.
Answer D is wrong. The gear has helical teeth.

Question #42
Answer A is wrong. The teeth should not be rounded.
Answer B is correct.
Answer C is wrong. Only Technician B is right.
Answer D is wrong. Technician B is right.

Question #43
Answer A is wrong. The front prop shaft must be disconnected from the transfer case to remove the case.
Answer B is correct. Most transfer cases can be removed without removing the vehicle transmission.
Answer C is wrong. The rear prop shaft must be disconnected from the transfer case to remove the case.
Answer D is wrong. Linkage must be disconnected from the transfer case.

Question #44
Answer A is wrong. A misaligned engine and transmission would cause the input shaft to put undue stress on the pilot bearing or bushing.
Answer B is wrong. A bent input shaft or clutch disc could cause a misalignment between the engine and transmission.
Answer C is correct. Both Technicians A and B are right.
Answer D is wrong. Both Technicians A and B are right.

Question #45
Answer A is wrong. Technician A is right; most transfer cases are assembled with RTV sealant.
Answer B is wrong. Technician B is right; some transfer cases are assembled with paper gaskets.
Answer C is correct. Both Technicians A and B are correct.
Answer D is wrong. Both Technicians A and B are correct.

Question #46
Answer A is wrong. The transaxle case must be opened before the input shaft can be removed.
Answer B is wrong. The transaxle case must be opened before the output shaft can be removed.
Answer C is correct. The pilot bearing is not part of the transaxle.
Answer D is wrong. The transaxle case must be opened before the differential bearings can be removed.

Question #47
Answer A is wrong. 10W-30 oil cannot be used on a limited slip differential.
Answer B is correct.
Answer C is wrong. Only Technician B is right.
Answer D is wrong. Technician B is right.

Question #48
Answer A is wrong. The best way to check for wear is to measure clearances.
Answer B is wrong. The best way to check for wear is to measure clearances.
Answer C is wrong. The best way to check for wear is to measure clearances.
Answer D is correct.

Question #49
Answer A is correct. The collapsible spacer must always be replaced.
Answer B is wrong. Pinion depth is adjusted using the collapsible spacer. There is no such thing as a "selective race."
Answer C is wrong. Only Technician A is right.
Answer D is wrong. Technician A is right.

Question #50
Answer A is wrong. A cork gasket is used to seal components without additional sealant.
Answer B is wrong. A rubber gasket is used to seal components without additional sealant.
Answer C is correct. Both Technicians A and B are right.
Answer D is wrong. Both Technicians A and B are right.

Question #51
Answer A is wrong. A worn blocking ring could cause the fluid to contain gold-colored material.
Answer B is correct. A worn second gear would produce metal shavings in the fluid.
Answer C is wrong. A worn thrust washer could cause the fluid to contain gold-colored material.
Answer D is wrong. A worn shift fork could cause the fluid to contain gold-colored material.

Question #52
Answer A is wrong. These gears turn when the vehicle turns.
Answer B is wrong. These gears are shimmed for proper fit and clearance.
Answer C is correct. Both Technicians A and B are right.
Answer D is wrong. Both Technicians A and B are right.

Question #53
Answer A is wrong. The axle shafts must be removed to remove the differential assembly.
Answer B is wrong. The bearing caps should always be marked to the housing to ensure proper installation.
Answer C is wrong. The shim packs and bearing races should be kept in order.
Answer D is correct. On some vehicles the pinion and differential assembly are removed as an assembly.

Question #54
Answer A is wrong. If the measurement was taken with the axle fully assembled, the torque of the whole assembly would be measured.
Answer B is wrong. The axle shafts do not need to be removed to measure pinion bearing preload.
Answer C is correct. Only the pinion bearing preload torque is wanted.
Answer D is wrong. An inch-pound torque wrench with a needle-type indicator should be used to measure pinion bearing preload.

Question #55
Answer A is correct. Dry or sticking linkage can cause hard shifting.
Answer B is wrong. Too strong a pressure plate would not cause hard shifting.
Answer C is wrong. Only Technician A is right.
Answer D is wrong. Technician A is right.

Question #56
Answer A is correct. Releases bearing noise will be present with the pedal depressed.
Answer B is wrong. A worn input shaft bearing will not cause a chirping sound with the clutch pedal depressed.
Answer C is wrong. Only Technician A is right.
Answer D is wrong. Technician A is right.

Question #57
Answer A is wrong. The driven gear is located on the vehicle speed sensor.
Answer B is wrong. The driving gear is located on the output shaft.
Answer C is correct.
Answer D is wrong. The speedometer driving gear is not located on the axle shaft.

Question #58
Answer A is wrong. The axle must be rotated slowly to see the changes in the runout.
Answer B is wrong. The runout is checked with a dial indicator.
Answer C is correct. Both Technicians A and B are right.
Answer D is wrong. Both Technicians A and B are right.

Question #59
Answer A is wrong. Worn input shaft splines could cause rough clutch engagement.
Answer B is wrong. Worn torsional springs may cause rough clutch engagement.
Answer C is correct. A worn pilot bearing is least likely to cause clutch chatter.
Answer D is wrong. A bent clutch disc could cause rough clutch engagement.

Question #60
Answer A is wrong. Lug nuts must be removed to remove the wheels for access to the front axles.
Answer B is wrong. If the vehicle has drum brakes on the front wheels, the drums must be removed for access to the front axles.
Answer C is correct. Brake pads or shoes do not need to be removed to check axle end play.
Answer D is wrong. Wheels must be removed for access to the axles.

Question #61
Answer A is wrong. The yoke seal should keep fluid from appearing at the yoke opening.
Answer B is wrong. A transfer case does not have a sight glass for checking fluid levels.
Answer C is correct.
Answer D is wrong. A piece of mechanic's wire inserted into the fill hole is not a practiced level-checking method.

Question #62
Answer A is wrong. Axle bearings have no relation to the ring-and-pinion contact pattern.
Answer B is correct. A worn collapsible spacer for the drive pinion will affect the ring-and-pinion contact pattern.
Answer C is wrong. Spider gears have no relation to the ring-and-pinion contact pattern.
Answer D is wrong. Differential bearings do not normally affect the ring-and-pinion contact pattern.

Question #63
Answer A is correct. A drive shaft normally is balanced before installation.
Answer B is wrong.
Answer C is wrong.
Answer D is wrong.

Question #64
Answer A is correct.
Answer B is wrong. A thinner shim should be used if the turning torque is more than specified.
Answer C is wrong. A thinner shim should be used if the turning torque is more than specified.
Answer D is wrong. The thicker shim should only be installed in the bell housing side of the transfer case.

Question #65
Answer A is wrong. The procedure does not indicate shift cable removal.
Answer B is correct.
Answer C is wrong. Shift cables should never be modified.
Answer D is wrong. The transaxle case is not being repaired.

Question #66
Answer A is wrong. 0.005 inch (0.127 mm) is less than minimum.
Answer B is wrong. 0.008 inch (0.203 mm) is less than minimum.
Answer C is correct.
Answer D is wrong. 0.025 inch (0.638 mm) is more than minimum.

Question #67
Answer A is wrong. A shift fork is connected to the synchronizer assembly.
Answer B is wrong. A shift fork is not connected to the counter shaft.
Answer C is wrong. A shift fork is not connected to the blocker rings.
Answer D is correct.

Question #68
Answer A is wrong. Mounts saturated in oil should never be cleaned and reused.
Answer B is wrong. Mounts soaked in oil should always be replaced.
Answer C is wrong. Oil will cause the mounts to swell and weaken.
Answer D is correct.

Question #69
Answer A is wrong. Stripped drive and driven gears could cause an inoperative speedometer.
Answer B is wrong. If the drive gear slips on the end of the sensor, the speedometer may be inoperative.
Answer C is correct. The drive gear position is not adjustable.
Answer D is wrong. If the driven gear teeth are stripped, the speedometer will be inoperative.

Question #70
Answer A is wrong. A hydraulic clutch has its own master cylinder.
Answer B is wrong. The slave cylinder is connected to the clutch master cylinder by hydraulic tubing and hose.
Answer C is wrong. Both Technicians A and B are wrong.
Answer D is correct. Both Technicians A and B are wrong.

Question #71
Answer A is correct.
Answer B is wrong. Snap rings and spacers can be obtained in a small kit.
Answer C is wrong. Only Technician A is right.
Answer D is wrong. Technician A is right.

Question #72
Answer A is correct. Spider gears rotate during cornering.
Answer B is wrong. Spider gears do not rotate at different speeds when the car is going straight.
Answer C is wrong. Only Technician A is right.
Answer D is wrong. Technician A is right

Question #73
Answer A is wrong. When the left-side adjuster nut is loosened, the backlash is increased.
Answer B is wrong. Pinion preload is adjusted with a collapsible spacer.
Answer C is wrong. Both Technicians A and B are wrong.
Answer D is correct. Both Technicians A and B are wrong.

Question #74
Answer A is wrong. SAE 90 gear oil is not usually used in late-model manual transmissions.
Answer B is wrong. Power steering fluid should not be used in a transmission.
Answer C is correct. Late-model manual transmissions use automatic transmission fluid or motor oil.
Answer D is wrong. Such a lubricant would be highly uncommon.

Question #75
Answer A is wrong. This bearing is usually sealed and does not require maintenance.
Answer B is correct.
Answer C is wrong. This bearing is only used on special applications.
Answer D is wrong. The support bearing is serviceable with the drive shaft removed.

Question #76
Answer A is wrong. Ring gear runout is not being measured.
Answer B is correct.
Answer C is wrong. Ring gear backlash is being measured.
Answer D is wrong. Bearing preload is not being measured.

Question #77
Answer A is wrong. Axle seals are not made of friction material.
Answer B is wrong. Pinion seals are not made of friction material.
Answer C is correct.
Answer D is wrong. Gaskets are not made of friction material.

Question #78
Answer A is wrong. Although sealers and gaskets do a good job of sealing the mating surfaces, the surfaces must be clean and free of burrs, or leakage may occur.
Answer B is correct.
Answer C is wrong. Only Technician B is right.
Answer D is wrong. Technician B is right.

Question #79
Answer A is wrong.
Answer B is correct. The wheel bearing is least likely to be inspected when replacing a CV boot.
Answer C is wrong.
Answer D is wrong.

Question #80
Answer A is wrong.
Answer B is wrong.
Answer C is wrong.
Answer D is correct. A dial indicator is used to check ring gear runout.

Question #81
Answer A is correct.
Answer B is wrong. Gear problems or tire problems would likely cause a loud humming noise.
Answer C is wrong. The pinion gear is not as likely to be affected by an excessive driveline angle.
Answer D is wrong. Transmission mount damage is not as likely to result from an excessive driveline angle.

Question #82
Answer A is wrong. A rusted shift linkage would cause hard shifting at all times.
Answer B is wrong. If the clutch disc was sticking to the flywheel, the hard shifting would occur at all times.
Answer C is correct.
Answer D is wrong. A bad slave cylinder would cause hard shifting at all times.

Question #83
Answer A is wrong.
Answer B is wrong.
Answer C is wrong.
Answer D is correct. Worn spider gears are the least likely to cause growling on deceleration.

Question #84
Answer A is wrong.
Answer B is wrong.
Answer C is correct. The differential side bearings are pressed on to the differential and therefore held in place by an interference fit.
Answer D is wrong.

Question #85
Answer A is correct.
Answer B is wrong. Gear journals are not indicated.
Answer C is wrong. Oil journals are not indicated.
Answer D is wrong. Synchronizer mounting locations are not indicated.

Question #86
Answer A is wrong.
Answer B is wrong.
Answer C is wrong.
Answer D is correct. Pinion bearing preload is measured with a torque wrench on the pinion nut.

Question #87
Answer A is correct.
Answer B is wrong. The spider gears do not contain needle bearings.
Answer C is wrong. Only Technician A is right.
Answer D is wrong. Technician A is right.

Question #88
Answer A is wrong. The center support bearing will not make noise if the vehicle is stopped.
Answer B is correct.
Answer C is wrong. Only Technician B is right.
Answer D is wrong. Technician B is right.

Question #89
Answer A is wrong. A helicoil would be the correct way to repair damaged threads in a transaxle.
Answer B is wrong. Some, not all, threads in a transaxle can be repaired. See the appropriate service manual.
Answer C is correct. Both Technicians A and B are right.
Answer D is wrong. Both Technicians A and B are right.

Question #90
Answer A is wrong.
Answer B is wrong.
Answer C is correct. A feeler gauge is used to measure blocking ring to gear face clearance.
Answer D is wrong.

Question #91
Answer A is correct.
Answer B is wrong. A damaged gear must always be replaced.
Answer C is wrong. Only Technician A is right.
Answer D is wrong. Technician A is right.

Question #92
Answer A is wrong. Some component that has contact with the shaft must have worn the shaft.
Answer B is wrong. If the shaft is worn, the case and all related components must be checked.
Answer C is correct.
Answer D is wrong. Both Technicians A and B are right.

Question #93
Answer A is wrong. The drive shaft must be removed to check pinion flange runout.
Answer B is wrong. A dial indicator would be used to check pinion flange runout.
Answer C is correct.
Answer D is wrong. Pinion flange runout cannot be checked with the flange removed.

Question #94
Answer A is correct. The reading on the dial indicator must be subtracted from 0.100 inch to obtain the normal pinion depth shim thickness. Therefore, the nominal shim thickness is 0.043 inch and the pinion marking of –4 is subtracted from this figure.
Answer B is wrong. A 0.041 inch (1.04 mm) shim would be too thick.
Answer C is wrong. A 0.042 inch (1.07 mm) shim would be too thick.
Answer D is wrong. A 0.043 inch (1.09 mm) shim would be too thick.

Question #95
Answer A is correct. Friction plates have a minimum thickness and should be measured with a micrometer.
Answer B is wrong. A feeler gauge could only measure the clearance between plates.
Answer C is wrong. Visual inspection is not accurate enough for measurement purposes.
Answer D is wrong. The friction plates must be removed from the clutch pack to measure specific plates.

Question #96
Answer A is wrong. Torsional dampening springs should face the pressure plate.
Answer B is correct.
Answer C is wrong. Only Technician B is right.
Answer D is wrong. Technician B is right.

Question #97
Answer A is wrong.
Answer B is wrong.
Answer C is correct. Both Technicians A and B are right.
Answer D is wrong. Both Technicians A and B are right.

Question #98
Answer A is correct.
Answer B is wrong. An oil-soaked transmission mount should be replaced.
Answer C is wrong. Only Technician A is right.
Answer D is wrong. Technician A is right.

Question #99
Answer A is wrong. The output shaft seal is not the least likely component that may need to be replaced.
Answer B is wrong. The tail housing gasket is not the least likely component that may need to be replaced.
Answer C is wrong. The speedometer O-ring is not the least likely component that may need to be replaced.
Answer D is correct. The counter gear shaft is housed in the transmission case, not the extension housing.

Question #100
Answer A is wrong. Input shaft end play should be checked before disassembling the transaxle.
Answer B is correct.
Answer C is wrong. Only Technician A is right.
Answer D is wrong. Technician B is right.

Question #101
Answer A is wrong. The cable is the shift linkage and may have stretched or frayed.
Answer B is wrong. The bushings and grommets may have become worn and need to be replaced.
Answer C is correct. Both Technicians A and B are right.
Answer D is wrong. Both Technicians A and B are right.

Question #102
Answer A is wrong. If the pinion gear is slipping on the ring gear in the axle housing, it would cause harsh noises other than clicking. The vehicle probably would not be driveable.
Answer B is correct. A stripped ring gear would make significantly more noise than just clicking, as well as causing other driveline problems.
Answer C is wrong. Only Technician B is right.
Answer D is wrong. Technician B is right.

Question #103
Answer A is wrong. The ring gear does not contact the clutch.
Answer B is correct.
Answer C is wrong. Only Technician B is right.
Answer D is wrong. Technician B is right.

Question #104
Answer A is wrong. A loose pinion nut can affect ring-and-pinion contact and produce a clunk.
Answer B is wrong. A loose pinion nut can increase pinion end play in the bearings and accelerate wear.
Answer C is wrong. A loose pinion nut can affect ring-and-pinion contact and produce the wrong wear pattern.
Answer D is correct. Spider gears have no relation to the tightness of the pinion nut.

Question #105
Answer A is wrong. A new mainshaft is not necessary to set end play.
Answer B is wrong. Bearings may be reused if they are inspected and in satisfactory condition.
Answer C is correct. Select-fit shims and thrust washers are used to set parts to specification.
Answer D is wrong. Snap rings may be reused if they are not damaged, and snap rings are not used to set this end play.

Question #106
Answer A is wrong. Drive shaft runout is checked with a dial indicator positioned near the center of the drive shaft.
Answer B is wrong. Drive shaft runout is checked with a dial indicator positioned near the center of the drive shaft.
Answer C is wrong. Both Technicians A and B are wrong.
Answer D is correct. Both Technicians A and B are wrong.

Question #107
Answer A is correct. Bearings are always lubricated before installation.
Answer B is wrong. Bearings are not installed with a horizontal slide press, and they are not installed dry.
Answer C is wrong. Only Technician A is right.
Answer D is wrong. Technician A is right.

Question #108
Answer A is wrong. Metallic material should be checked for during a transmission oil change.
Answer B is wrong. Type of oil and condition should be checked during a transmission oil change.
Answer C is wrong. Leaks should be checked during a transmission oil change.
Answer D is correct. Manual transaxles do not have filters.

Question #109
Answer A is correct.
Answer B is wrong. A worn blocker ring would have dull edges in the cone area.
Answer C is wrong. Only Technician A is right.
Answer D is wrong. Technician A is right.

Question #110
Answer A is wrong. Tapered bearings would not be used on an input shaft.
Answer B is wrong. Needle bearings could not support the weight of the input shaft.
Answer C is wrong. The input bearing is lubricated by transmission fluid.
Answer D is correct. A ball bearing assembly is commonly used on an input shaft.

Question #111
Answer A is correct. A straightedge is the best way to check the mating surfaces.
Answer B is wrong.
Answer C is wrong.
Answer D is wrong.

Question #112
Answer A is wrong. A bad electronic shift motor may cause the transfer case to not engage.
Answer B is wrong. A blown fuse may cause the transfer case to not engage.
Answer C is wrong. A bad 4WD engage switch may cause the transfer case to not engage.
Answer D is correct. An electronic transfer case does not have any linkage exposed that can rust.

Question #113
Answer A is wrong. Transfer cases do not have sight glasses.
Answer B is wrong. Transfer cases do not have fill vents.
Answer C is wrong. Both Technicians A and B are wrong.
Answer D is correct. Both Technicians A and B are wrong.

Question #114
Answer A is correct. An axle shaft should be stored standing up.
Answer B is wrong.
Answer C is wrong. Runout cannot be checked in this way.
Answer D is wrong.

Question #115
Answer A is wrong. Improper adjustment may not let the clutch fully engage.
Answer B is wrong. Improper adjustment may not disengage the clutch, causing the gears to grind.
Answer C is correct. Both Technicians A and B are right.
Answer D is wrong. Both Technicians A and B are right.

Question #116
Answer A is wrong. Measurements for tolerances and end play should be recorded and transferred to the transmission upon rebuilding it.
Answer B is wrong. All parts should be cleaned with solvent.
Answer C is wrong. Always pay attention to the condition of parts being removed; this may help in identifying the source of the problem.
Answer D is correct. Friction discs are not part of a manual transmission.

Question #117
Answer A is wrong. The gear may not be making a good contact with the gear on the output shaft.
Answer B is wrong. A broken cable may catch and slip repeatedly; or a bad head may be sticking or slipping.
Answer C is correct. Both Technicians A and B are right.
Answer D is wrong. Both Technicians A and B are right.

Question #118
Answer A is wrong. Third gear may cause this wear.
Answer B is wrong. Fourth gear may cause this wear.
Answer C is wrong. The third and fourth blocking rings may cause this wear.
Answer D is correct.

Question #119
Answer A is wrong. The drive shaft should be inspected for damage before balancing.
Answer B is wrong. Screw-type hose clamps can be used for balance weights.
Answer C is wrong. A strobe light should be used to view the chalk marks on the drive shaft.
Answer D is correct. The suspension should be supported at its normal ride height.

Question #120
Answer A is wrong. Worn friction plates may cause the limited slip differential to not function properly.
Answer B is wrong. Weak spring tension could cause a limited slip differential to fail.
Answer C is correct. The friction plates are not under excessive load and are unlikely to strip the teeth.
Answer D is wrong. The wrong fluid in the differential could cause a limited slip differential to fail.

Question #121
Answer A is wrong. Differential side bearings do not need to be packed with grease.
Answer B is wrong. A hydraulic press should be used to install the differential side bearings.
Answer C is correct. If the bearing race is reused, premature bearing failure may occur.
Answer D is wrong. The case should only be replaced if there are signs of damage.

Question #122
Answer A is wrong. Starter alignment may cause this problem.
Answer B is wrong. Missing engine dowels may cause this problem.
Answer C is wrong. A damaged or rubbing inspection cover may cause this problem.
Answer D is correct. A worn transmission main shaft bearing would not cause a grinding noise when trying to start the engine.

Question #123
Answer A is wrong. A bad front axle shaft would not cause a clicking noise while turning.
Answer B is wrong. A constant-velocity inner joint would be affected when the vehicle's suspension rebounds.
Answer C is wrong. A bad torsional damper would cause a shudder in the vehicle, not a clicking sound.
Answer D is correct. The ball bearings in the outer CV joint are probably damaged.

Question #124
Answer A is wrong. The shift motor drives the shift linkages that are housed inside the transfer case.
Answer B is wrong. An open would not power or signal the transfer case motor to shift the unit into 4WD.
Answer C is correct. Both Technicians A and B are right.
Answer D is wrong. Both Technicians A and B are right.

Question #125
Answer A is wrong. A bent linkage may prevent the transmission from engaging reverse.
Answer B is wrong. A broken shift fork may prevent the transmission from engaging reverse.
Answer C is wrong. A misadjusted linkage may prevent the transmission from engaging reverse.
Answer D is correct. The transmission would engage in reverse, but may function poorly.

Question #126
Answer A is wrong.
Answer B is wrong.
Answer C is wrong.
Answer D is correct. The spider gears are only replaced if they are worn.

Question #127
Answer A is wrong. The input shaft bearing should be checked.
Answer B is wrong. The input shaft bearing retainer and seal should be checked.
Answer C is wrong. The crankshaft should be checked.
Answer D is correct. The output shaft bushing does not contact the pilot bushing.

Question #128
Answer A is wrong. The flywheel should be resurfaced every time the clutch assembly is replaced.
Answer B is correct.
Answer C is wrong. Only Technician B is right.
Answer D is wrong. Technician B is right.

Question #129
Answer A is wrong.
Answer B is wrong.
Answer C is correct. Both Technicians A and B are right.
Answer D is wrong. Both Technicians A and B are right.

Question #130
Answer A is wrong.
Answer B is wrong.
Answer C is correct. U-joints are most likely to be damaged by excessive drive shaft angle.
Answer D is wrong.

Question #131
Answer A is wrong. Some differential noises can be mistaken for other vehicle noises.
Answer B is wrong. The differential gears only turn when the vehicle is moving.
Answer C is correct. Both Technicians A and B are right.
Answer D is wrong. Both Technicians A and B are right.

Question #132
Answer A is wrong. The clutch would not be making even contact with the flywheel.
Answer B is wrong. Oil would contaminate the clutch disc and cause uneven clutch disc wear.
Answer C is correct. Both Technicians A and B are right.
Answer D is wrong. Both Technicians A and B are right.

Question #133
Answer A is correct.
Answer B is wrong. The reverse idler gear does not make contact in first therefore it would not make any noise in first.
Answer C is wrong. Only Technician A is right.
Answer D is wrong. Technician A is right.

Question #134
Answer A is correct.
Answer B is wrong. The bent shift fork should be discarded and a new one installed.
Answer C is wrong. Only Technician A is right.
Answer D is wrong. Technician A is right.

Question #135
Answer A is correct. A sensor should never be replaced as the first step after reading a trouble code
Answer B is wrong. The harness could be tested as the next step after reading a trouble code.
Answer C is wrong. The service manual could be consulted as the next step after reading a trouble code.
Answer D is wrong. The connector could be inspected as the next step after reading a trouble code.

Question #136
Answer A is wrong. The ring gear tooth pattern has nothing to do with carrier bearing preload.
Answer B is correct.
Answer C is wrong. Only Technician B is right.
Answer D is wrong. Technician B is right.

Question #137
Answer A is wrong.
Answer B is wrong.
Answer C is correct. The differential side bearings would be the most likely cause.
Answer D is wrong.

Question #138
Answer A is correct.
Answer B is wrong. The pressure plate should be replaced not resurfaced.
Answer C is wrong. Only Technician A is right.
Answer D is wrong. Technician A is right.

Glossary

Abrasion Wearing or rubbing away of a part.

Acceleration An increase in velocity or speed.

Adhesives Chemicals used to hold gaskets in place during the assembly of an engine. They also aid the gasket in maintaining a tight seal by filling in the small irregularities on the surfaces and by preventing the gasket from shifting due to engine vibration.

Alignment An adjustment to a line or to bring into a line.

Antifriction bearing A bearing designed to reduce friction. This type of bearing normally uses ball or roller inserts to reduce the friction.

Anti-seize Thread compound designed to keep threaded connections from damage due to rust or corrosion.

Automatic locking/unlocking hubs Front wheel hubs that can engage or disengage themselves from the axles automatically.

Axial Parallel to a shaft or bearing bore.

Axis The centerline of a rotating part, a symmetrical part, or a circular bore.

Axle The shaft or shafts of a machine upon which the wheels are mounted.

Axle carrier assembly A cast-iron framework that can be removed from the rear axle housing for service and adjustment of the parts.

Axle housing Designed in the removable carrier or integral carrier types to house the drive pinion, ring gear, differential, and axle shaft assemblies.

Axle shaft A shaft on which the road wheels are mounted.

Axle shaft end thrust A force exerted on the end of an axle shaft that is most pronounced when the vehicle turns corners and curves.

Axle shaft tube A tube attached to the axle housing center section to surround the axle shaft and bearings.

Backlash The amount of clearance or play between two meshed gears.

Balance Having equal weight distribution. The term is usually used to describe the weight distribution around the circumference and between the front and back sides of a wheel and tire assembly.

Ball bearing An antifriction bearing that consists of a hardened inner and outer race with hardened steel balls that roll between the two races and support the load of the shaft.

Ball-and-trunnion universal joint A non–constant-velocity universal joint that combines the universal joint and slip joint.

Ball joint A suspension component that attaches the control arm to the steering knuckle and serves as the lower pivot point for the steering knuckle.

Bearing The supporting part that reduces friction between a stationary and rotating part or between two moving parts.

Bearing caps In the differential, caps held in place by bolts or nuts that, in turn, hold bearings in place.

Bearing cone The inner race, rollers, and cage assembly of a tapered-roller bearing which must always be replaced in matched sets.

Bearing cup The outer race of a tapered-roller bearing or ball bearing.

Bearing race The surface on which the rollers or balls of a bearing rotate. The outer race is the same thing as the cup, and the inner race is the one closest to the axle shaft.

Belleville spring A tempered spring steel cone-shaped plate used to aid the mechanical force in a pressure plate assembly.

Bell housing A housing that fits over the clutch components and connects the engine and the transmission.

Bellows Rubber protective covers with accordion-like pleats used to contain lubricants and keep out contaminating dirt or water.

Bevel spur gear Gear that has teeth with a straight centerline cut on a cone.

Boot A term used for bellows.

Brinnelling Rough lines worn across a bearing race or shaft due to impact loading, vibration, or inadequate lubrication.

Burr A feather edge of metal left on a part being cut with a file or other cutting tool.

Bushing A cylindrical lining used as a bearing assembly; made of steel, brass, bronze, nylon, or plastic.

C-clip A C-shaped clip used to retain the driveaxles in some rear axle assemblies.

Canceling angles Opposing operating angles of two universal joints cancel the vibrations developed by the individual universal joint.

Cardan universal joint A non–constant-velocity universal joint consisting of two yokes with their forked ends joined by a cross. The driven yoke changes speed twice in 360 degrees of rotation.

Castellated nut A nut with six raised portions or notches through which a cotter pin can be inserted to secure the nut.

Center hanger bearing Ball-type bearing mounted on a vehicle crossmember to support the driveshaft and provide better installation angle to the rear axle.

Center section The middle of the integral axle housing containing the drive pinion, ring gear, and differential assembly.

Centering joint Ball socket joint placed between two Cardan universal joints to ensure that the assembly rotates on center.

Centrifugal clutch A clutch that uses centrifugal force to apply a higher force against the friction disc as the clutch spins faster.

Chamfer A bevel or taper at the edge of a hole or a gear tooth.

Chamfer face A beveled surface on a shaft or part that allows for easier assembly.

Chase To straighten up or repair damaged threads.

Chassis The vehicle frame, suspension, and running gear.

Chuckle A rattling noise that sounds much like a stick rubbing against the spokes of a bicycle wheel.

Circlip A split steel snap ring that fits into a groove to hold various parts in place.

Clamp bolt Another name for a pinch bolt.

Clashing Grinding sound heard when gear and shaft speeds are not the same during a gearshift operation.

Clearance The space allowed between two parts, such as between a journal and a bearing.

Close ratio A relative term for describing the gear ratios in a transmission.

Clunking A metallic noise most often heard when a transmission is engaged in reverse or drive, caused by excessive backlash somewhere in the driveline and felt or heard in the axle.

Cluster assembly A manual transmission related term applied to a group of gears of different sizes machined from one steel casting.

Clutch A device for connecting and disconnecting the engine from the transmission or for a similar purpose in other units.

Clutch control cable A cable assembly with a flexible outer housing anchored at the upper and lower ends with an inside cable that transfers clutch pedal movement to the clutch release lever.

Clutch (friction) disc The friction material part of the clutch assembly that fits between the flywheel and pressure plate.

Clutch fork A Y-shaped member into which the throwout bearing is assembled.

Clutch housing A large aluminum or iron casting that surrounds the clutch assembly, located between the engine and transmission.

Clutch linkage A combination of shafts, levers, or cables that transmits clutch pedal motion to the clutch assembly.

Clutch packs A series of clutch discs and plates installed alternately in a housing to act as a driving or driven unit.

Clutch pedal A pedal in the driver's compartment that operates the clutch.

Clutch push rod A solid or hollow rod that transfers linear motion between movable parts; that is, the clutch release bearing and release plate.

Clutch shaft A term used for the transmission input shaft or main drive pinion. The clutch driven disc drives this shaft.

Clutch slippage A condition whereby engine speed increases but increased torque is not transferred through to the driving wheels.

Clutch teeth The locking teeth of a gear.

Coil spring A heavy wire-like steel coil used to support the vehicle weight while allowing for suspension motions.

Coil preload springs Coil springs that are made of tempered steel rods formed into a spiral that resist compression; located in the pressure plate assembly.

Coil spring clutch A clutch using coil springs to hold the pressure plate against the friction disc.

Companion flange A mounting flange that fixedly attaches a driveshaft to another drive train component.

Cone clutch The driving and driven parts conically shaped to connect and disconnect power flow. A clutch made from two cones, one fitting inside the other. Friction between the cones forces them to rotate together.

Constant mesh Manual transmission design permits the gears to be constantly enmeshed regardless of vehicle operating circumstances.

Constant mesh transmission A transmission in which the gears are engaged at all times, and shifts are made by sliding collars, clutches, or other means to connect the gears to the output shaft.

Constant-velocity joint (CV joint) A flexible coupling between two shafts that permits each shaft to maintain the same driving or driven speed regardless of operating angle, allowing for a smooth transfer of power.

Cotter pin A type of fastener, made from soft steel in the form of a split pin, that can be inserted in a drilled hole. The split ends are spread to lock the pin in position.

Counterclockwise (ccw) rotation Rotating the opposite direction of the hands on a clock.

Countergear assembly A cluster of gears designed on one casting with short shafts supported by antifriction bearings.

Countershaft An intermediate shaft that receives motion from a mainshaft and transmits it to a working part, sometimes called a lay shaft.

Coupling A connecting means for transferring movement from one part to another; may be mechanical, hydraulic, or electrical.

Coupling yoke A part of the double Cardan universal joint that connects the two universal joint assemblies.

Cover plate A stamped steel cover bolted over the service access to the manual transmission.

Crossmember A steel part of the frame structure that transverses the vehicle body to connect the longitudinal frame rails. Crossmembers can be welded into place or removed from the vehicle.

CV joint Constant-velocity joint.

Dead axle An axle that only supports the vehicle and does not transmit power.

Detent A small depression in a shaft, rail, or rod into which a pawl or ball drops when the shaft, rail, or rod is moved, providing a locking effect.

Dial indicator A measuring instrument with the readings indicated on a dial rather than on a thimble as on a micrometer.

Diaphragm spring A circular disc shaped like a cone, with spring tension that allows it to flex forward or backward.

Diaphragm spring clutch A clutch in which a diaphragm spring, rather than a coil spring, applies pressure against the friction disc.

Differential A mechanism between driveaxles that permits one wheel to run at a different speed than the other while turning.

Differential action An operational situation where one driving wheel rotates at a slower speed than the opposite driving wheel.

Differential case The metal unit that encases the differential side gears and pinion gears, and to which the ring gear is attached.

Differential case spread Another name for preload.

Differential drive gear A large circular helical gear driven by the transaxle pinion gear and shaft and which drives the differential assembly.

Differential housing Cast-iron assembly that houses the differential unit and the driveaxles. This is also called the rear axle housing.

Differential pinion gears Small beveled gears located on the differential pinion shaft.

Differential pinion shaft A short shaft locked to the differential case. This shaft supports the differential pinion gears.

Differential ring gear A large circular hypoid-type gear enmeshed with the hypoid drive pinion gear.

Differential side gears The gears inside the differential case that are internally splined to the axle shafts, and are driven by the differential pinion gears.

Direct drive One turn of the input driving member compared to one complete turn of the driven member, such as when there is direct engagement between the engine and driveshaft where the engine crankshaft and the driveshaft turn at the same rpm.

Disengage When the operator moves the clutch pedal toward the floor to disconnect the driven clutch disc from the driving flywheel and pressure plate assembly.

Dog tooth A series of gear teeth that are part of the dog clutching action in a transmission synchronizer operation. The locking teeth of a gear.

Double Cardan universal joint A near constant-velocity universal joint that consists of two Cardan universal joints connected by a coupling yoke.

Double-offset constant-velocity joint Another name for the type of plunging, inner CV joint found on many GM, Ford, and Japanese FWD cars.

Double reduction axle A driveaxle construction in which two sets of reduction gears are used for extreme reduction of the gear ratio.

Dowel pin A pin inserted in matching holes in two parts to maintain those parts in fixed relation to one another.

Downshift To shift a transmission into a lower gear.

Driveline The universal joints, driveshaft, and other parts connecting the transmission with the driving axles.

Driveline torque Relates to rear wheel driveline and is the transfer of torque between the transmission and the driving axle assembly.

Driveline wrapup A condition where axles, gears, U-joints, and other components can bind or fail if the 4WD mode is used on pavement where 2WD is more suitable.

Drive pinion The gear that takes its power directly from the driveshaft or transmission and drives the ring gear.

Drive pinion flange A rim used to connect the rear of the driveshaft to the rear axle drive pinion.

Drive pinion gear One of the two main driving gears located within the transaxle or rear driving axle housing. Together the two gears multiply engine torque.

Driveshaft An assembly of one or two universal joints connected to a shaft or tube; used to transmit power from the transmission to the differential. Also called the propeller shaft.

Driveshaft installation angle The angle the driveshaft is mounted off the true horizontal line; measured in degrees.

Driven disc The part of the clutch assembly that receives driving motion from the flywheel and pressure plate assemblies.

Driven gear The gear meshed directly with the driving gear to provide torque multiplication, reduction, or a change of direction.

Driving axle A term related collectively to the rear driving axle assembly where the drive pinion, ring gear, and differential assembly are located within the driving axle housing.

Dry-disc clutch A clutch in which the friction faces of the friction disc are dry, as opposed to a wet-disc clutch, which runs submerged in oil. The conventional type of automobile clutch.

Dual-reduction axle A driveaxle construction with two sets of pinions and gears, either of which can be used.

Dummy shaft A shaft, shorter than the countershaft, used during disassembly and reassembly in place of the countershaft.

Dynamic balance The balance of an object when it is in motion; for example, the dynamic balance of a rotating driveshaft.

Eccentric One circle within another circle wherein both circles do not have the same center or one circle is mounted off center.

Eccentric washer A normal looking washer with its hole not in its center. The hole is offset from the center.

Emulsification When water droplets mix with grease resulting in a thicker solution than normal grease.

End play The amount of axial or end-to-end movement in a shaft due to clearance in the bearings.

Engage When the vehicle operator moves the clutch pedal up from the floor, this engages the driving flywheel and pressure plate to rotate and drive the driven disc.

Engagement chatter A shaking, shuddering action that takes place as the driven disc makes contact with the driving members. Chatter is caused by a rapid grip and slip action.

Engine The source of power for most vehicles. It converts burned fuel energy into mechanical force.

Extension housing An aluminum or iron casting of various lengths that encloses the transmission output shaft and supporting bearings.

External cone clutch The external surface of one part has a tapered surface to mate with an internally tapered surface to form a cone clutch.

External gear A gear with teeth across the outside surface.

Externally tabbed clutch plates Clutch plates are designed with tabs around the outside periphery to fit into grooves in a housing or drum.

Extreme-pressure lubricant A special lubricant for use in hypoid gear differentials; needed because of the heavy wiping loads imposed on the gear teeth.

Final drive gears Main driving gears located in the axle area of the transaxle housing.

Final drive ratio The ratio between the drive pinion and ring gear.

First gear A small diameter driving helical- or spur-type gear located on the cluster gear assembly. First gear provides torque multiplication to get the vehicle moving.

Fit The contact between two machined surfaces.

Fixed-type constant-velocity joint A joint that cannot telescope or plunge to compensate for suspension travel. Fixed joints are always found on the outer ends of the driveshafts of FWD cars. A fixed joint may be of either Rzeppa or tripod type.

Flange yoke The part of the rear universal joint attached to the drive pinion.

Fluid coupling A device in the powertrain consisting of two rotating members; transmits power from the engine, through a fluid, to the transmission.

Fluid drive A drive in which there is no mechanical connection between the input and output shafts and power is transmitted by moving oil.

Flywheel A heavy metal wheel that is attached to the crankshaft and rotates with it; helps smooth out the power surges from the engine power strokes; also serves as part of the clutch and engine-cranking system.

Flywheel ring gear A gear, fitted around the flywheel, that is engaged by teeth on the starting-motor drive to crank the engine.

Forward coast side The side of the ring gear tooth the drive pinion contacts when the vehicle is decelerating.

Forward drive side The side of the ring gear tooth that the drive pinion contacts when accelerating or on the drive.

Four-wheel drive (4WD) A vehicle system with driving axles at both front and rear, so that all four wheels can be driven. 4WD is the standard abbreviation for four-wheel drive.

Four-wheel high A transfer case shift position where both front and rear driveshafts receive power and rotate at the speed of the transmission output shaft.

Frame The main understructure of the vehicle to which everything else is attached. Most FWD cars have only a subframe for the front suspension and drive train. The body serves as the frame for the rear suspension.

Free-wheeling clutch A mechanical device that will engage the driving member to impart motion to a driven member in one direction but not the other. Also known as an overrunning clutch.

Friction bearing A bearing in which there is sliding contact between the moving surfaces. Sleeve bearings, such as those used in connecting rods, are friction bearings.

Friction disc In the clutch, a flat disc faced on both sides with friction material and splined to the clutch shaft. It is positioned between the clutch pressure plate and the engine flywheel. Also called the clutch disc or driven disc.

Friction facings A hard-molded or woven asbestos or paper material that is riveted or bonded to the clutch driven disc.

Front bearing retainer An iron or aluminum circular casting fastened to the front of a transmission housing to retain the front transmission bearing assembly.

Front differential/axle assembly Like a conventional rear axle but having steerable wheels.

Front-wheel drive (FWD) The vehicle has all drive train components located at the front.

Fulcrum rings A circular ring over which the pressure plate diaphragm spring pivots.

Full-floating rear axle An axle that only transmits driving force to the rear wheels. The weight of the vehicle (including payload) is supported by the axle housing.

Fully synchronized In a manual transmission, the synchronizer assembly operates to improve the shift quality in all forward gears.

Galling Wear caused by metal-to-metal contact in the absence of adequate lubrication. Metal is transferred from one surface to the other, leaving behind a pitted or scaled appearance.

Gasket A layer of material, usually made of cork, paper, plastic, composition, or metal, or a combination of these, placed between two parts to make a tight seal.

Gasket cement A liquid adhesive material, or sealer, used to install gaskets.

Gear A wheel with external or internal teeth that serves to transmit or change motion.

Gear clash The noise that results when two gears are traveling at different speeds and are forced together.

Gear lubricant A type of grease or oil blended especially to lubricate gears.

Gear noise The howling or whining of the ring gear and pinion due to an improperly set gear pattern, gear damage, or improper bearing preload.

Gear ratio The number of revolutions of a driving gear required to turn a driven gear through one complete revolution. For a pair of gears, the ratio is found by dividing the number of teeth on the driven gear by the number of teeth on the driving gear.

Gear rattle A repetitive metallic impact or rapping noise that occurs when the vehicle is lugging in gear.

Gear reduction When a small gear drives a large gear, there is an output speed reduction and a torque increase that results in a gear reduction.

Gear whine A high-pitched sound developed by some types of meshing gears.

Gearshift A linkage-type mechanism by which the gears in an automobile transmission are engaged and disengaged.

Half shaft Either of the two driveshafts that connect the transaxle to the wheel hubs in FWD vehicles, may be of solid or tubular steel and may be of different lengths.

Harshness A bumpy ride caused by a stiff suspension. Can often be cured by installing softer springs or shock absorbers.

Helical gear Gears with the teeth cut at an angle to the axis of the gear.

Herringbone gear A pair of helical gears designed to operate together. The angle of the pair of gears forms a V.

High pedal A clutch pedal that has an excessive amount of pedal travel.

Hot spots The small areas on a friction surface that are a different color, normally blue, or are harder than the rest of the surface.

Hotchkiss drive A type of rear suspension in which leaf springs absorb the rear axle housing torque.

Hub The center part of a wheel, to which the wheel is attached.

Hydraulic clutch A clutch that is actuated by hydraulic pressure; used in cars and trucks when the engine is some distance from the driver's compartment so that it would be difficult to use mechanical linkages.

Hypoid gear A gear that is similar in appearance to a spiral bevel gear, but the teeth are cut so that the gears match in a position where the shaft centerlines do not meet; cut in a spiral form to allow the pinion to be set below the centerline of the ring gear so that the car floor can be lower.

Hypoid gear lubricant An extreme pressure lubricant designed for the severe operating conditions of hypoid gears.

ID Inside diameter.

Inboard constant-velocity joint. The inner constant-velocity joint, or the one closest to the transaxle.

Inclinometer Device designed with a spirit level and graduated scale to measure the inclination of a driveline assembly.

Index To orient two parts by marking them. During reassembly, the parts are arranged so the index marks are next to each other. Used to preserve the orientation between balanced parts.

Input shaft The shaft carrying the driving gear by which the power is applied, as to the transmission.

Inserts One of several terms that could apply to the shift plates found in a synchronizer assembly.

Insert springs Round wire springs that hold the inserts or shift plates in contact with the synchronizer sleeve, located around the synchronizer hub.

Integral axle housing. A rear axle housing-type where the parts are serviced through an inspection cover and adjusted within and relative to the axle housing.

Interference fit A press fit; for example, if the inside diameter of a bore is 0.001 inch (0.0254 mm) smaller than the outside diameter of a shaft, the shaft must be pressed in.

Interlock mechanism A mechanism in the transmission shift linkage that prevents the selection of two gears at one time.

Intermediate bearing plate Another name for the center support plate of a transmission.

Intermediate driveshaft Located between the left and right driveshafts, it equalizes driveshaft length.

Intermediate plate A mechanism in the transmission shift linkage that prevents the selection of two gears at one time.

Internal gear A gear with teeth pointing inward, toward the hollow center of the gear.

Joint angle The angle formed by the input and output shafts of constant-velocity joints.

Journal A bearing with a hole in it for a shaft.

Key A small block inserted between a shaft and hub to prevent circumferential movement.

Keyway A groove or slot cut to permit the insertion of a key.

Knock A heavy metallic sound usually caused by a loose or worn bearing.

Knurl To indent or roughen a finished surface.

Lash The amount of free motion in a gear train, between gears, or in a mechanical assembly, such as the lash in a valve train.

Leaf spring A spring made up of a single flat steel plate or of several plates of graduated lengths assembled one on top of another; used on vehicles to absorb road shocks by bending or flexing.

Limited slip differential A differential designed so that when one wheel is slipping, a major portion of the drive torque is supplied to the wheel with the better traction; also called a non-slip differential.

Linkage Any series of rods, yokes, and levers, etc., used to transmit motion from one unit to another.

Locked differential A differential with the side and pinion gears locked together.

Locking A condition of a bearing caused by large particles of dirt that become trapped between a bearing and its race.

Locknut A second nut turned down on a holding nut to prevent loosening.

Lockplates Metal tabs bent around nuts or bolt heads.

Lockwasher A type of washer that, when placed under the head of a bolt or nut, prevents the bolt or nut from working loose.

Low speed The gearing that produces the highest torque and lowest speed of the wheels.

Lubricant Any material, usually a petroleum product such as grease or oil, that is placed between two moving parts to reduce friction.

Lug nut The nuts that fasten the wheels to the axle hub or brake rotor. Missing lug nuts should always be replaced. Overtightening can cause warpage of the brake rotor in some cases.

Lugging A term used to describe an operating condition in which the engine is operating at too low of an engine speed for the selected gear.

Matched gearset code Identification marks on two gears that indicate they are matched. They should not be mismatched with another gearset and placed into operation.

Meshing The mating, or engaging, of the teeth of two gears.

Mounts Made of rubber to insulate vibrations and noise while they support a powertrain part, such as engine or transmission mounts.

Multiple disc A clutch with a number of driving and driven discs as compared to a single plate clutch.

Needle bearing An antifriction bearing using a great number of long, small-diameter rollers.

Neutral In a transmission, the setting in which all gears are disengaged and the output shaft is disconnected from the drive wheels.

Neutral start switch A switch wired into the ignition switch to prevent engine cranking unless the transmission shift lever is in neutral or the clutch pedal is depressed.

Noise Any unwanted or annoying sound.

Nominal shim A shim with a designated thickness.

Nut A removable fastener used with a bolt to lock pieces together; made by threading a hole through the center of a piece of metal that has been shaped to a standard size.

O-ring A type of sealing ring, usually made of rubber or a rubberlike material. In use, the O-ring is compressed into a groove to provide the sealing action.

OD Outside diameter.

Oil seal A seal placed around a rotating shaft or other moving part to prevent leakage of oil.

One-way clutch A term used for sprag clutch.

Outboard constant-velocity joint The constant-velocity joint closest to the wheels.

Output shaft The shaft or gear that delivers the power from a device, such as a transmission.

Overcenter spring A heavy coil spring arrangement in the clutch linkage to assist the driver with disengaging the clutch and returning the clutch linkage to the full engagement position.

Overdrive Any arrangement of gearing that produces more revolutions of the driven shaft than of the driving shaft.

Overdrive ratio Identified by the decimal point indicating less than one driving input revolution compared to one output revolution of a shaft.

Overrun coupling A free-wheeling device to permit rotation in one direction but not in the other.

Overrunning clutch A device consisting of a shaft or housing linked together by rollers or sprags operating between movable and fixed races.

Pawl A lever that pivots on a shaft. When lifted it swings freely and when lowered it locates in a detent or notch to hold a mechanism stationary.

Pedal play The distance the clutch pedal and release bearing assembly move from the fully engaged position to the point where the release bearing contacts the pressure plate release levers.

Phasing Rotational position of the universal joints on the driveshaft.

Pilot bearing A small bearing, such as in the center of the flywheel end of the crankshaft, which carries the forward end of the clutch shaft.

Pilot bushing A plain bearing fitted in the end of a crankshaft. The primary purpose is to support the input shaft of the transmission.

Pilot shaft A shaft used to align parts and that is removed before final installation of the parts; a dummy shaft.

Pinion gear The smaller of two meshing gears.

Pinion carrier The mounting or bracket that retains the bearings supporting a pinion shaft.

Pivot A pin or shaft upon which another part rests or turns.

Planet carrier In a planetary gear system, the carrier or bracket in a planetary system that contains the shafts upon which the pinions or planet gears turn.

Planet gears The gears in a planetary gearset that connect the sungear to the ring gear.

Planet pinions In a planetary gear system, the gears that mesh with, and revolve about, the sungear; they also mesh with the ring gear.

Planetary gearset A system of gearing modeled after the solar system. A pinion is surrounded by an internal ring gear and planet gears are in mesh between the ring gear and pinion around which all revolve.

Plate loading Force developed by the pressure plate assembly to hold the driven disc against the flywheel.

Plunging action Telescoping action of an inner front-wheel-drive universal joint.

Plunging constant-velocity joint Usually the inner constant-velocity joint. The joint is designed so that it can telescope slightly to compensate for suspension motions.

Powertrain The mechanisms that carry the power from the engine crankshaft to the drive wheels; these include the clutch, transmission, driveline, differential, and axles.

Preload A load applied to a part during assembly so as to maintain critical tolerances when the operating load is applied later.

Press fit Forcing a part into an opening that is slightly smaller than the part itself to make a solid fit.

Pressure plate That part of the clutch that exerts force against the friction disc; it is mounted on and rotates with the flywheel. A heavy steel ring pressed against the clutch disc by spring pressure.

Propeller shaft A term used for driveshaft.

Pulsation To move or beat with rhythmic impulses.

Quadrant A section of a gear. A term sometimes used to identify the shift lever selector mounted on the steering column.

Quill shaft The term used by some manufacturers to refer to the protruding hollow shaft of the transmission's front bearing retainer.

Race A channel in the inner or outer ring of an antifriction bearing in which the balls or rollers roll.

Raceway A groove or track designed into the races of a bearing or universal joint housing to guide and control the action of the balls or trunnions.

Radial The direction moving straight out from the center of a circle. Perpendicular to the shaft or bearing bore.

Radial clearance (radial displacement) Clearance within the bearing and between balls and races perpendicular to the shaft.

Radial load A force perpendicular to the axis of rotation.

Ratcheting mechanism Uses a pawl and gear arrangement to transmit motion or to lock a particular mechanism by having the pawl drop between gear teeth.

Ravigneaux Designer of a planetary gear system with small and large sungears, long and short planetary pinions, planetary carriers, and a ring gear.

Rear axle torque The torque received and multiplied by the rear driving axle assembly.

Rear-wheel drive (RWD) A term associated with a vehicle where the engine is mounted at the front and the driving axle and driving wheels at the rear of the vehicle.

Release bearing A ball-type bearing moved by the clutch pedal linkage to contact the pressure plate release levers to either engage or disengage the driven disc with the clutch driving members.

Release levers In the clutch, levers that are moved by throwout bearing movement, causing clutch spring force to be relieved so that the clutch is disengaged, or uncoupled, from the flywheel.

Release plate Plate designed to release the clutch pressure plate's loading on the clutch driven disc.

Removable carrier housing A type of rear axle housing from which the axle carrier assembly can be removed for parts service and adjustment.

Retaining ring A removable fastener used as a shoulder to retain and position a round bearing in a hole.

Retractor clips Spring steel clips that connect the diaphragm's flexing action to the pressure plate.

Reverse idler gear In a transmission, an additional gear that must be meshed to obtain reverse gear; a gear used only in reverse that does not transmit power when the transmission is in any other position.

Ring gear A gear that surrounds or rings the sun and planet gears in a planetary system. Also the name given to the spiral bevel gear in a differential.

Roller bearing Any tube to form a gasket of any shape.

Rubber coupling Rubber-based disc used as a universal joint between the driving and driven shafts.

Runout Deviation of the specified normal travel of an object. The amount of deviation or wobble a shaft or wheel has as it rotates.

Rzeppa constant-velocity joint The name given to the ball-type constant-velocity joint (as opposed to the tripod-type constant-velocity joint). Rzeppa joints are usually the outer joints on most FWD cars. Named after its inventor, Alfred Rzeppa, a Ford engineer.

Safety stands Commonly called jack stands; used to support a vehicle when it is raised by a jack or hoist.

Score A scratch, ridge, or groove marring a finished surface.

Scuffing A type of wear in which there is a transfer of material between parts moving against each other; shows up as pits or grooves in the mating surfaces.

Seal A material, shaped around a shaft, used to close off the operating compartment of the shaft, preventing oil leakage.

Sealer A thick, tacky compound, usually spread with a brush, which may be used as a gasket or sealant to seal small openings or surface irregularities.

Seat A surface, usually machined, upon which another part rests or seats; for example, the surface upon which a valve face rests.

Self-adjusting clutch linkage Monitors clutch pedal play through a clutch control cable and ratcheting mechanism to automatically adjust clutch pedal play.

Semicentrifugal pressure plate The release levers of this pressure plate are weighted to take advantage of centrifugal force to increase plate loading resulting in reduced driven disc slip.

Semifloating rear axle An axle that supports the weight of the vehicle on the axle shaft in addition to transmitting driving forces to the rear wheels.

Shift forks Mechanisms attached to shift rails that fit into the synchronizer hub for change of gears.

Shift lever The lever used to change gears in a transmission. Also, the lever on the starting motor that moves the drive pinion into or out of mesh with the flywheel teeth.

Shift rails Rods placed within the transmission housing that are a part of the transmission gearshift linkage.

Shim Thin sheets used as spacers between two parts, such as the two halves of a journal bearing.

Shudder A shake or shiver movement.

Side gears Gears that are meshed with the differential pinions and splined to the axle shafts (RWD) or driveshafts (FWD).

Side thrust Longitudinal movement of two gears.

Slave cylinder Located at a lower part of the clutch housing. Receives fluid pressure from the master cylinder to engage or disengage the clutch.

Sliding fit Where sufficient clearance has been allowed between the shaft and journal to allow free-running without overheating.

Sliding yoke Slides on internal and external splines to compensate for driveline length changes.

Sliding gear transmission A transmission in which gears are moved on their shafts to change gear ratios.

Slip fit Running or sliding fit.

Slipjoint In the powertrain, a variable-length connection that permits the driveshaft to change its effective length.

Snap ring Split spring-type ring located in an internal or external groove to retain a part.

Solid axle A rear axle design that places the final drive, axles, bearings, and hubs into one housing.

Spalling A condition of a bearing that is caused by overloading the bearing and is evident by pits on the bearings or their races.

Speed gears Driven gears located on the transmission output shaft. This term differentiates between the gears of the countergear and cluster assemblies and gears on the transmission output shaft.

Spindle The shaft on which the wheels and wheel bearings mount.

Spiral bevel gear A ring gear and pinion wherein the mating teeth are curved and placed at an angle with the pinion shaft.

Spiral gear A gear with teeth cut according to a mathematical curve on a cone. Spiral bevel gears that are not parallel have centerlines that intersect.

Spline Slot or groove cut in a shaft or bore; a splined shaft onto which a hub, wheel, gear, etc., with matching splines in its bore, is assembled so that the two must turn together.

Splined hub Several keys placed radially around the inside diameter of a circular part, such as a wheel or driven disc.

Split lip seal Typically a rope seal sometimes used to denote any two-part oil seal.

Split pin A round split spring steel tubular pin used for locking purposes; for example, locking a gear to a shaft.

Sprag clutch A member of the overrunning clutch family using a sprag to jam between the inner and outer races used for holding or driving action.

Spring A device that changes shape when it is stretched or compressed, but returns to its original shape when the force is removed; the component of the automotive suspension system that absorbs road shocks by flexing and twisting.

Spring retainer A steel plate designed to hold a coil or several coil springs in place.

Spur gear Gears cut on a cylinder with teeth that are straight and parallel to the axis.

Squeak A high-pitched noise of short duration.

Squeal A continuous high-pitched noise.

Stabilizer bar Also called a sway bar. It prevents the vehicle's body from diving into turns.

Strut assembly Refers to all the strut components, including the strut tube, shock absorber, coil spring, and upper bearing assembly.

Stub shaft A very short shaft.

Sungear The central gear in a planetary gear system around which the rest of the gears rotate. The innermost gear of the planetary gearset.

Sway bar Also called a stabilizer bar. It prevents the vehicle's body from diving into turns.

Synchromesh transmission Transmission gearing that aids the meshing of two gears or shift collars by matching their speed before engaging them.

Synchronize To cause two events to occur at the same time; for example, to bring two gears to the same speed before they are meshed in order to prevent gear clash.

Synchronizer assembly Device that uses cone clutches to bring two parts rotating at two speeds to the same speed. A synchronizer assembly operates between two gears; e.g., first and second gear, third and fourth gear.

Synchronizer blocker ring Usually a brass ring that acts as a clutch and causes driving and driven units to turn at the same speed before final engagement.

Synchronizer hub Center part of the synchronizer assembly that is splined to the synchronizer sleeve and transmission output shaft.

Synchronizer sleeve The sliding sleeve that fits over the complete synchronizer assembly.

Tail shaft A commonly used term for a transmission's extension housing.

Three-quarter floating axle The axle housing carries the weight of the vehicle while the bearings support the wheels on the outer ends of the axle housing tubes.

Throwout bearing In the clutch, the bearing that can be moved inward to the release levers by clutch-pedal action to cause declutching, which disengages the engine crankshaft from the transmission.

Thrust washer A washer designed to take up end thrust and prevent excessive end play.

Tolerance A permissible variation between the two extremes of a specification or dimension.

Torque A twisting motion, usually measured in ft.-lbs. (N•m).

Torque multiplication The result of meshing a small driving gear and a large driven gear to reduce speed and increase output torque.

Torque steer An action felt in the steering wheel as the result of increased torque.

Torque tube A fixed tube over the driveshaft on some cars. It helps locate the rear axle and takes torque reaction loads from the driveaxle so the driveshaft will not sense them.

Torsional springs Round, stiff coil springs placed in the driven disc to absorb the torsional disturbances between the driving flywheel and pressure plate and the driven transmission input shaft.

Total pedal travel The total amount the pedal moves from no free play to complete clutch disengagement.

Total travel Distance the clutch pedal and release bearing move from the fully engaged position until the clutch is fully disengaged.

Traction The gripping action between the tire tread and the road's surface.

Transaxle Type of construction in which the transmission and differential are combined in one unit.

Transaxle assembly A compact housing most often used in front-wheel-drive vehicles that houses the manual transmission, final drive gears, and differential assembly.

Transfer case An auxiliary transmission mounted behind the main transmission. Used to divide engine power and transfer it to both front and rear differentials, either full time or part time.

Transmission The device in the powertrain that provides different gear ratios between the engine and drive wheels as well as reverse.

Transmission case An aluminum or iron casting that encloses the manual transmission parts.

Transverse Powertrain layout in a front-wheel-drive automobile extending from side to side.

Tripod (also called tripot) A three-prong bearing that is the major component in tripod constant-velocity joints. It has three arms (or trunnions) with needle bearings and rollers that ride in the grooves or yokes of a tulip assembly.

Tripod universal joints Universal joint consisting of a hub with three arm and roller assemblies that fit inside a casting called a tulip.

Trunnion One of the projecting arms on a tripod or on the cross of a four-point universal joint. Each trunnion has a bearing surface that allows it to pivot within a joint or slide within a tulip assembly.

Tulip assembly The outer housing containing grooves or yokes in which trunnion bearings move in a tripod constant-velocity joint.

Two-disk clutch A clutch with two friction discs for additional holding power; used in heavy-duty equipment.

Two-speed rear axle A term used for a double-reduction differential.

U-bolt An iron rod with threads on both ends, bent into the shape of a U and fitted with a nut at each end.

U-joint A four-point cross connected to two U-shaped yokes that serves as a flexible coupling between shafts.

Universal joint A mechanical device that transmits rotary motion from one shaft to another shaft at varying angles.

Universal joint operating angle The difference in degrees between the driveshaft and transmission installation angles.

Unsprung weight The weight of the tires, wheels, axles, control arms, and springs.

Upshift To shift a transmission into a higher gear.

Vibration A quivering, trembling motion felt in the vehicle at different speed ranges.

Viscous friction The friction between layers of a liquid.

Wet-disc clutch A clutch in which the friction disc (or discs) is operated in a bath of oil.

Wheel A disc or spokes with a hub at the center that revolves around an axle, with a rim around the outside for mounting the tire on.

Wheel offset The amount of the wheel assembly that is to the side of the wheel's mounting hub.

Wheel shimmy The wobble of a tire.

Worm gear A gear with teeth that resemble a thread on a bolt. It is meshed with a gear that has teeth similar to a helical tooth except that it is dished to allow more contact.

Yoke In a universal joint, the driveable torque-and-motion input and output member, attached to a shaft or tube.

Yoke bearing A U-shaped, spring-loaded bearing in the rack-and-pinion steering assembly that presses the pinion gear against the rack.